STATISTICS
HACKS™

Other resources from O'Reilly

Related titles Baseball Hacks™   Access Hacks™
Mind Hacks™   Excel Hacks™
Mind Performance Hacks™ Online Investing Hacks™
Astronomy Hacks™

Hacks Series Home *hacks.oreilly.com* is a community site for developers and power users of all stripes. Readers learn from each other as they share their favorite tips and tools for Mac OS X, Linux, Google, Windows XP, and more.

oreilly.com *oreilly.com* is more than a complete catalog of O'Reilly books. You'll also find links to news, events, articles, weblogs, sample chapters, and code examples.

oreillynet.com is the essential portal for developers interested in open and emerging technologies, including new platforms, programming languages, and operating systems.

Conferences O'Reilly brings diverse innovators together to nurture the ideas that spark revolutionary industries. We specialize in documenting the latest tools and systems, translating the innovator's knowledge into useful skills for those in the trenches. Visit *conferences.oreilly.com* for our upcoming events.

Safari Bookshelf (*safari.oreilly.com*) is the premier online reference library for programmers and IT professionals. Conduct searches across more than 1,000 books. Subscribers can zero in on answers to time-critical questions in a matter of seconds. Read the books on your Bookshelf from cover to cover or simply flip to the page you need. Try it today.

STATISTICS
HACKS™

Bruce Frey

O'REILLY®

Beijing · Cambridge · Farnham · Köln · Paris · Sebastopol · Taipei · Tokyo

Statistics Hacks™
by Bruce Frey

Published by O'Reilly Media, Inc., 1005 Gravenstein Highway North,
Sebastopol, CA 95472.

O'Reilly books may be purchased for educational, business, or sales promotional use. Online editions are also available for most titles (*safari.oreilly.com*). For more information, contact our corporate/institutional sales department: (800) 998-9938 or *corporate@oreilly.com*.

Editor: Brian Sawyer
Production Editor: Genevieve d'Entremont
Copyeditor: Genevieve d'Entremont
Proofreader: Mary Brady

Indexer: Lucie Haskins
Cover Designer: Linda Palo
Interior Designer: David Futato
Illustrators: Robert Romano and Jessamyn Read

Printing History:

May 2006: First Edition.

 This book uses RepKover™, a durable and flexible lay-flat binding.

ISBN: 0-596-10164-3
[M]   [5/06]

Contents

Credits

About the Author

Bruce Frey, Ph.D., is a comic book collector and film buff. In his spare time, he teaches statistics to graduate students and conducts research in his secret identity as an assistant professor in Educational Psychology and Research at the University of Kansas. He is an award-winning teacher, and his scholarly research interests are in the areas of teacher-made tests and classroom assessment, the measurement of spirituality, and program evaluation methods. Bruce's honors include taking third place in the Kansas Monopoly Championship as a teenager, second place in the Kansas Film Festival as a college student, and a respectable third-place finish in the Lawrence, Kansas, Texas Hold 'Em Poker Tournament as a middle-aged man. He is proudest of two accomplishments: his marriage to his sweet wife, and his purchase of a low-grade copy of *Showcase* #4, a comic book wherein the "Silver Age Flash first appears," whatever that means.

Contributors

The following people contributed their hacks, writing, and inspiration to this book:

- Joseph Adler is the author of *Baseball Hacks* (O'Reilly), and a researcher in the Advanced Product Development Group at VeriSign, focusing on problems in user authentication, managed security services, and RFID security. Joe has years of experience analyzing data, building statistical models, and formulating business strategies as an employee and consultant for companies including DoubleClick, American Express, and Dun & Bradstreet. He is a graduate of the Massachusetts Institute of Technology with an Sc.B. and an M.Eng. in computer science and computer engineering. Joe is an unapologetic Yankees fan, but he appreciates any good baseball game. Joe lives in Silicon Valley with his wife, two cats, and a DirecTV satellite dish.

- Ron Hale-Evans is a writer, thinker, and game designer who earns his daily sandwich with frequent gigs as a technical writer. He has a Bachelor's degree in Psychology from Yale, with a minor in Philosophy. Thinking a lot about thinking led him to create the Mentat Wiki (*http://www.ludism.org/mentat*), which led to his recent book, *Mind Performance Hacks* (O'Reilly). You can find his multinefarious [sic] other projects at his home page, *http://ron.ludism.org*, including his award-winning board games, a list of his Short-Duration Personal Saviors, and his blog. Ron's next book will probably be about game systems, especially since his series of articles on that topic for the dearly departed *The Games Journal* (*http://www.thegamesjournal.com*) has been relatively popular among both gamers and academics. If you want to email Ron the names of some gullible publishers, or if you just want to bug him, you can reach him at *rwhe@ludism.org* (rhymes with *nudism* and has nothing to do with *Luddism*).

- Brian E. Hansen, 27, grew up in the Dallas, Texas area. After serving a two-year religious mission in Spain, he attended Texas A&M University and graduated in 2004 with a B.S. degree in Petroleum Engineering. He currently works as a Reservoir Engineer for a large independent oil and gas exploration and production company headquartered in Irving, Texas.

- Jill H. Lohmeier received her Ph.D. in Cognitive Psychology from The University of Massachusetts, Amherst. She is currently the Evaluation Director for the School Program Evaluation and Research group at the University of Kansas. Jill likes outdoor sports, especially running, hiking, and playing soccer with her kids.

- Ernest E. Rothman is a Professor and Chair of the Mathematical Sciences Department at Salve Regina University (SRU) in Newport, Rhode Island. Ernie holds a Ph.D. in Applied Mathematics from Brown University and held positions at the Cornell Theory Center in Ithaca, New York before coming to SRU. His interests are primarily in scientific computing, mathematics and statistics education, and the Unix underpinnings of Mac OS X. You can keep abreast of his latest activities at *http://homepage.mac.com/samchops*.

- Neil J. Salkind is a sometimes faculty member at the University of Kansas with an office opposite that of Bruce Frey, of *Statistics Hacks* fame. In addition to being the author of *Statistics for People Who (Think They) Hate Statistics* (SAGE), Neil is a developmental psychologist who collects books, cooks, works on old houses and a p1800 Volvo, and is active in Masters swimming. He has also written over 100 trade books and textbooks, and works with StudioB Literary Agency in New York.

- William Skorupski is currently an assistant professor in the School of Education at the University of Kansas, where he teaches courses in psychometrics and statistics. He earned his Bachelor's degree in educational research and psychology from Bucknell University in 2000, and his Doctorate in psychometric methods from the University of Massachusetts, Amherst in 2004. His primary research interest is in the application of mathematical models to psychometric data, including the use of Bayesian statistics for solving practical measurement problems. He also enjoys applying his knowledge of statistics and probability to everyday situations, such as playing poker against the author of this book!

Acknowledgments

I'd like to thank all the contributors to this book, both those who are listed in the "Contributors" section and those who helped with ideas, reviewed the manuscript, and provided suggestions of sources and resources. Thanks in this capacity especially go to Tim Langdon, neon bender, whose gift of Harry Blackstone, Jr.'s paperback book *There's One Born Every Minute* (Jove Publications) provided great inspiration for many of the hacks herein.

I'd like to thank my editor, Brian Sawyer, who shepherded this project with a strong hand and a strong vision of what is and is not a *hack*. He was right most of the time. (Though not all the time, Brian. That hack about using a monkey to pick the winner of the Kentucky Derby should have made it in. Maybe next time....) Brian was instrumental in bringing this project to completion, especially during a string of unlucky rolls where the odds of success looked slim.

I'd like to thank Neil Salkind, statistics writer supreme, for his help with many facets of my professional life and this book.

Most importantly, thanks to Bonnie Johnson, my sweet wife, whom I vaguely recall, but who I think will be waiting for me at home when I finally turn in the last revision of this book.

Preface

Chance plays a huge part in your life, whether you know it or not. Your particular genetic makeup mutated slightly when you were created, and it did so based on specific laws of probability. Performance in school involves human errors, yours and others', which tends to keep your actual ability level from being reflected precisely in your report card or on those high-stakes tests. Research on careers even suggests that what you do for a living was probably not a result of careful planning and preparation, but more likely due to happenstance. And, of course, chance determines your fate in games of *chance* and plays a large role in the outcome of sporting events.

Fortunately, an entire set of scientific tools, the various applications of statistics, can be used to solve the problems caused by our fate-influenced system. Inferential statistics, a field of science based entirely on the nature of probability, allows us to understand the way things work, discover relationships among variables, describe a huge population by seeing just a small bit of it, make uncannily accurate predictions, and, yes, even make a little money with a well-placed wager here and there.

This book is a collection of statistical tricks and tools. *Statistics Hacks* presents useful tools from statistics, of course, but also from the realms of educational and psychological measurement and experimental research design. It provides solutions to a variety of problems in the world of social science, but also in the worlds of business, games, and gambling.

If you are already a top scientist and do statistical calculations in your sleep, you'll enjoy this book and the creative applications it finds for those rusty old tools you know so well. If you just like the scientific approach to life and are entertained by cool ideas and clever solutions to interesting problems, don't worry. *Statistics Hacks* was written with the nonscientist in mind, too, so if that is you, you've come to the right place. It's written for the nonstatistician as well, so if this still describes you, you'll feel safe here.

If, on the other hand, you are taking a statistics course or have some interest in the academic nature of the topic, you might find this book a pleasant companion to the textbooks typically required for those sorts of courses. There won't be any contradictions between your textbook and this book, so hearing about real-world applications of statistical tools that seem only theoretical won't hurt your development. It's just that there are some pretty cool things that you can do with statistics that seem more like fun than like work.

Why Statistics Hacks?

The term *hacking* has a bad reputation in the press. They use it to refer to people who break into systems or wreak havoc, using computers as their weapon. Among people who write code, though, the term *hack* refers to a "quick-and-dirty" solution to a problem or a clever way to get something done. And the term *hacker* is taken very much as a compliment, referring to someone as being *creative*, having the technical chops to get things done. The Hacks series is an attempt to reclaim the word, document the good ways people are hacking, and pass the hacker ethic of creative participation on to the uninitiated. Seeing how others approach systems and problems is often the quickest way to learn about a new technology.

The technologies at the heart of this book are statistics, measurement, and research design. Computer technology has developed hand-in-hand with these technologies, so the use of the term *hacks* to describe what is done in this book is consistent with almost every perspective on that word. Though there is just a little computer hacking covered in these pages, there is a plethora of *clever ways to get things done*.

How This Book Is Organized

You can read this book from cover to cover if you like, but each hack stands on its own, so feel free to browse and jump to the different sections that interest you most. If there's a prerequisite you need to know about, a cross-reference will guide you to the right hack.

The earlier hacks are more foundational and probably provide generalized solutions or strategic approaches across a variety of problems to a greater extent than later hacks. On the other hand, later hacks provide much more specific tricks for winning games or just information to help you understand what's going on around you.

The book is divided into several chapters, organized by subject:

Chapter 1, *The Basics*

Use these hacks as a strong set of foundational tools, the ones you will use most often when you are stat-hacking your way into and out of trouble. Think of these as your basic toolkit: your hammer, saw, and various screwdrivers.

Chapter 2, *Discovering Relationships*

This chapter covers statistical ways to find, describe, and test relationships among variables. You will be able to make the invisible visible with these hacks.

Chapter 3, *Measuring the World*

A variety of tips and tricks for measuring the world around you are presented here. You'll learn to ask the right questions, assess accurately, and even increase your own performance on high-stakes tests.

Chapter 4, *Beating the Odds*

This chapter is for the gambler. Use the odds to your advantage, and make the right decisions in Texas Hold 'Em poker and just about every other game in which probability determines the outcome.

Chapter 5, *Playing Games*

From TV game show strategy to winning Monopoly to enjoying sports to just having fun, this chapter presents different hacks for getting the most out of your game playing.

Chapter 6, *Thinking Smart*

This chapter is perhaps the most cerebral of them all. Get your mind right, play mind games, make discoveries, and unlock the mysteries of the world around us using the statistics hacks you'll find here.

Conventions Used in This Book

The following is a list of the typographical conventions used in this book:

Italics

Used to indicate key terms and concepts, URLs, and filenames.

`Constant width`

Used for Excel functions and code examples.

`Constant width italic`

Used for code text that should be replaced by user-supplied values.

Gray type

Used to indicate a cross-reference within the text.

You should pay special attention to notes set apart from the text with this icon:

This is a tip, suggestion, or general note. It contains useful supplementary information about the topic at hand.

The thermometer icons, found next to each hack, indicate the relative complexity of the hack:

 beginner   moderate   expert

Safari® Enabled

 When you see a Safari® Enabled icon on the cover of your favorite technology book, that means the book is available online through the O'Reilly Network Safari Bookshelf.

Safari offers a solution that's better than e-books. It's a virtual library that lets you easily search thousands of top tech books, cut and paste code samples, download chapters, and find quick answers when you need the most accurate, current information. Try it for free at *http://safari.oreilly.com*.

How to Contact Us

We have tested and verified the information in this book to the best of our ability, but you may find that the rules or characteristics of a given situation are different than described here. As a reader of this book, you can help us to improve future editions by sending us your feedback. Please let us know about any errors, inaccuracies, misleading or confusing statements, and typos that you find anywhere in this book.

Please also let us know what we can do to make this book more useful to you. We take your comments seriously and will try to incorporate reasonable suggestions into future editions. You can write to us at:

O'Reilly Media, Inc.
1005 Gravenstein Hwy N.
Sebastopol, CA 95472
800-998-9938 (in the U.S. or Canada)
707-829-0515 (international/local)
707-829-0104 (fax)

To ask technical questions or to comment on the book, send email to:

bookquestions@oreilly.com

The web site for *Statistics Hacks* lists examples, errata, and plans for future editions. You can find this page at:

http://www.oreilly.com/catalog/statisticshks

For more information about this book and others, see the O'Reilly web site:

http://www.oreilly.com

Got a Hack?

To explore Hacks books online or to contribute a hack for future titles, visit:

http://hacks.oreilly.com

The Basics
Hacks 1–10

There's only a small group of tools that statisticians use to explore the world, answer questions, and solve problems. It is the way that statisticians use probability or knowledge of the normal distribution to help them out in different situations that varies. This chapter presents these basic hacks.

Taking known information about a distribution and expressing it as a probability [Hack #1] is an essential trick frequently used by stat-hackers, as is using a tiny bit of sample data to accurately describe all the scores in a larger population [Hack #2]. Knowledge of basic rules for calculating probabilities [Hack #3] is crucial, and you gotta know the logic of significance testing if you want to make statistically-based decisions [Hacks #4 and #8].

Minimizing errors in your guesses [Hack #5] and scores [Hack #6] and interpreting your data [Hack #7] correctly are key strategies that will help you get the most bang for your buck in a variety of situations. And successful stat-hackers have no trouble recognizing what the results of any organized set of observations or experimental manipulation really mean [Hacks #9 and #10].

Learn to use these core tools, and the later hacks will be a breeze to learn and master.

HACK #1 Know the Big Secret

Statisticians know one secret thing that makes them seem smarter than everybody else.

The primary purpose of statistics as a scientific methodology is to make probability statements about samples of scores. Before we jump into that, we need some quick definitions to get us rolling, both to understand this hack and to lay a foundation for other statistics hacks.

Samples are numeric values that you have gathered together and can see in front of you that represent some larger *population* of scores that you have not gathered together and cannot see in front of you. Because these values are almost always numbers that indicate the presence or level of some characteristic, measurement folks call these values *scores*. A *probability statement* is a statement about the likelihood of some event occurring.

Probability is the heart and soul of statistics. A common perception of statisticians, in fact, is that they mainly calculate the exact likelihood that certain events of interest will occur, such as winning the lottery or being struck by lightning. Historically, the person who had the tools to calculate the likely outcome of a dice game was the same person who had the tools to describe a large group of people using only a few summary statistics.

So, traditionally, the teaching of statistics includes at least some time spent on the basic rules of probability: the methods for calculating the chances of various combinations or permutations of possible outcomes. More common applications of statistics, however, are the use of *descriptive statistics* to describe a group of scores, or the use of *inferential statistics* to make guesses about a population of scores using only the information contained in a sample of scores. In social science, the scores usually describe either people or something that is happening to them.

It turns out, then, that researchers and measurers (the people who are most likely to use statistics in the real world) are called upon to do more than calculate the probability of certain combinations and permutations of interest. They are able to apply a wide variety of statistical procedures to answer questions of varying levels of complexity without once needing to compute the odds of throwing a pair of six-sided dice and getting three 7s in a row.

Those odds are .005 or 1/2 of 1 percent if you start from scratch. If you have already rolled two 7s, you have a 16.6 percent chance of rolling that third 7.

The Big Secret

The key reason that probability is so crucial to what statisticians do is because they like to make probability statements about the scores in real or theoretical distributions.

A *distribution* of scores is a list of all the different values and, sometimes, how many of each value there are.

For example, if you know that a quiz just administered in a class you are taking resulted in a distribution of scores in which 25 percent of the class got 10 points, then I might say, without knowing you or anything about you, that there is a 25 percent chance that you got 10 points. I could also say that there is a 75 percent chance that you did *not* get 10 points. All I have done is taken known information about the distribution of some values and expressed that information as a statement of probability. This is a trick. It is the secret trick that all statisticians know. In fact, this is mostly all that statisticians ever do!

Statisticians take known information about the distribution of some values and express that information as a statement of probability. This is worth repeating (or, technically, threepeating, as I first said it five sentences ago). Statisticians take *known information* about the distribution of some values and *express that information* as a statement of probability.

Heavens to Betsy, we can all do that. How hard could it be? Imagine that there are three marbles in an otherwise empty coffee can. Further imagine that you know that only one of the marbles is blue. There are three values in the distribution: one blue marble and two marbles of some other color, for a total sample size of three. There is one blue marble out of three marbles. Oh, statistician, what are the chances that, without looking, I will draw the blue marble out first? One out of three. 1/3. 33 percent.

To be fair, the values and their distributions most commonly used by statisticians are a bit more abstract or complex than those of the marbles in a coffee can scenario, and so much of what statisticians do is not quite that transparent. Applied social science researchers usually produce values that represent the difference between the average scores of several groups of people, for example, or an index of the size of the relationship between two or more sets of scores. The underlying process is the same as that used with the coffee can example, though: reference the known distribution of the value of interest and make a statement of probability about that value.

The key, of course, is how one *knows* the distribution of all these exotic types of values that might interest a statistician. How can one know the distribution of average differences or the distribution of the size of a relationship between two sets of variables? Conveniently, past researchers and mathematicians have developed or discovered formulas and theorems and rules of thumb and philosophies and assumptions that provide us with the knowledge of the distributions of these complex values most often sought by researchers. The work has been done for us.

A Smaller, Dirtier Secret

Most of the procedures that statisticians use to take known information about a distribution of scores and express that information as a statement of probability have certain requirements that must be met for the probability statement to be accurate. One of these assumptions that almost always must be met is that the values in a sample have been *randomly* drawn from the distribution.

Notice that in the coffee can example I slipped in that "without looking" business. If some force other than random chance is guiding the sampling process, then the associated probabilities reported are simply wrong and— here's the worst part—we can't possibly know how wrong they are. Much, and maybe most, of the applied psychological and educational research that occurs today uses samples of people that were not randomly drawn from some population of interest.

College students taking an introductory psychology course make up the samples of much psychological research, for example, and students at elementary schools conveniently located near where an educational researcher lives are often chosen for study. This is a problem that social science researchers live with or ignore or worry about, but, nevertheless, it is a limitation of much social science research.

HACK #2 Describe the World Using Just Two Numbers

Most of the statistical solutions and tools presented in this book work only because you can look at a sample and make accurate inferences about a larger population. The Central Limit Theorem is the meta-tool, the prime directive, the king of all secrets that allows us to pull off these inferential tricks.

Statistics provide solutions to problems whenever your goal is to describe a group of scores. Sometimes the whole group of scores you want to describe is in front of you. The tools for this task are called *descriptive statistics*. More often, you can see only part of the group of the scores you want to describe, but you still want to describe the whole group. This summary approach is called *inferential statistics*. In inferential statistics, the part of the group of scores you can see is called a *sample*, and the whole group of scores you wish to make inferences about is the *population*.

It is quite a trick, though, when you think about it, to be able to describe with any confidence a population of values when, by definition, you are not directly observing those values. By using three pieces of information—two sample values and an assumption about the shape of the distribution of scores in the population—you can confidently and accurately describe those invisible populations. The set of procedures for deriving that eerily accurate description is collectively known as the *Central Limit Theorem*.

Some Quick Statistics Basics

Inferential statistics tend to use two values to describe populations, the *mean* and the *standard deviation*.

Mean. Rather than describe a sample of values by showing them all, it is simply more efficient to report some fair summary of a group of scores instead of listing every single score. This single number is meant to fairly represent all the scores and what they have in common. Consequently, this single number is referred to as the *central tendency* of a group of scores.

Typically, the best measure of central tendency, for a variety of reasons, is the *mean* [Hack #21]. The mean is the arithmetic average of all the scores and is calculated by adding together all the values in a group, and then dividing that total by the number of values. The mean provides more information about all the scores in a group than other central tendency options (such as reporting the middle score, the most common score, and so on).

In fact, mathematically, the mean has an interesting property. A side effect of how it is created (adding up all scores and dividing by the number of scores) produces a number that is as close as possible to all the other scores. The mean will be close to some scores and far away from some others, but if you add up those distances, you get a total that is as small as possible. No other number, real or imagined, will produce a smaller total distance from all the scores in a group than the mean.

Standard deviation. Just knowing the mean of a distribution doesn't quite tell us enough. We also need to know something about the variability of the scores. Are they mostly close to the mean or mostly far from the mean? Two wildly different distributions could have the same mean but differ in their variability. The most commonly reported measure of variability summarizes the distances between each score and the mean.

As with the mean, the more informative measure of variability would be one that uses all the values in a distribution. A measure of variability that does this is the *standard deviation*. The standard deviation is the average distance of each score from the mean. A standard deviation calculates all the *distances* in a distribution and averages them. The "distances" referred to are the distance between each score and the mean.

Another commonly reported value that summarizes the variability in a distribution is the *variance*. The variance is simply the standard deviation squared and is not particularly useful in picturing a distribution, but it is helpful when comparing different distributions and is frequently used as a value in statistical calculations, such as with the independent *t test* [Hack #17].

The formula for the standard deviation appears to be more complicated than it needs to be, but there are some mathematical complications with summing distances (negative distances always cancel out the positive distances when the mean is used as the dividing point). Consequently, here is the equation:

$$\sqrt{\frac{\sum (x - \text{Mean})^2}{n - 1}}$$

Σ means to sum up. The x means each score, and the n means the number of scores.

Central Limit Theorem

The Central Limit Theorem is fairly brief, but very powerful. Behold the truth:

> If you randomly select multiple samples from a population, the means of each of those samples will be normally distributed.

Attached to the theorem are a couple of mathematical rules for accurately estimating the descriptive values for this imaginary distribution of sample means:

- The mean of these means (that's a mouthful) will be equal to the population mean. The mean of a single sample is a good estimate for this mean of means.

- The standard deviation of these means is equal to the sample standard deviation divided by the square root of the sample size, n:

$$\frac{\sigma}{\sqrt{n}}$$

These mathematical rules produce more accurate results, and the distribution is closer to the normal curve as the sample size within any sample gets bigger.

30 or more in a sample seems to be enough to produce accurate applications of the Central Limit Theorem.

So What?

Okay, so the Central Limit Theorem appears somewhat intellectually interesting and no doubt makes statisticians all giggly and wriggly, but what does it all mean? How can anyone *use* it to *do* anything cool?

As discussed in "Know the Big Secret" [Hack #1], the secret trick that all statisticians know is how to solve problems statistically by taking known information about the distribution of some values and expressing that information as a statement of probability. The key, of course, is how one *knows* the distribution of all these exotic types of values that might interest a statistician. How can one know the distribution of average differences or the distribution of the size of a relationship between two sets of variables? The Central Limit Theorem, that's how.

For example, to estimate the probability that any two groups would differ on some variable by a certain amount, we need to know the distribution of means in the population from which those samples were drawn. How could we possibly know what that distribution is when the population of means is invisible and might even be only theoretical? The Central Limit Theorem, Bub, that's how! How can we know the distributions of correlations (an index of the strength of a relationship between two variables) which could be drawn from a population of infinite possible correlations? Ever hear of the Central Limit Theorem, dude?

Because we know the proportion of values that reside all along the normal curve [Hack #23], and the Central Limit Theorem tells me that these summary values are normally distributed, I can place probabilities on each statistical outcome. I can use these probabilities to indicate the level of statistical significance (the level of certainty) I have in my conclusions and decisions. Without the Central Limit Theorem, I could hardly ever make statements about statistical significance. And what a drab, sad life that would be.

Applying the Central Limit Theorem

To apply the Central Limit Theorem, I need start with only a sample of values that I have randomly drawn from a population. Imagine, for example, that I have a group of eight new Cub Scouts. It's my job to teach them knot tying. I suspect, let's say, that this isn't the brightest bunch of Scouts who have ever come to me for knot-tying guidance.

Before I demand extra pay, I want to determine whether they are, in fact, a few badges short of a bushel. I want to know their IQ. I know that the population's average IQ is 100, but I notice that no one in my group has an intelligence test score above 100. I would expect at least some above that score. Could this group have been selected from that average population? Maybe my sample is just unusual and doesn't represent all Cubbies. A statistical approach, using the Central Limit Theorem, would be to ask:

> Is it possible that the mean IQ of the population represented by this sample is 100?

If I want to know something about the population from which my Scouts were drawn, I can use the Central Limit Theorem to pretty accurately estimate the population's mean IQ and its standard deviation. I can also figure out how much difference there is likely to be between the population's mean IQ and the mean IQ in my sample.

I need some data from my scouts to figure all this out. Table 1-1 should provide some good information.

Table 1-1. Scout smarts

Scout	IQ
Jimmy	100
Perry	95
Clark	90
Lex	92
Neil	85
Billy	88
Greg	93
John	91

The descriptive statistics for this sample of eight IQ scores are:

- Mean IQ = 91.75
- Standard deviation = 4.53

So, I know in my sample that most scores are within about $4^1/_2$ IQ points of 91.75. It is the invisible population they came from, though, that I am most interested in. The Central Limit Theorem allows me to estimate the population's mean, standard deviation, and, most importantly, how far sample means will likely stray from the population mean:

Mean IQ
Our sample mean is our best estimate, so the population mean is likely close to 91.75.

Standard deviation of IQ scores in the population

> The formula we used to calculate our sample standard deviation is designed especially to estimate the population standard deviation, so we'll guess 4.53.

Standard deviation of the mean

> This is the real value of interest. We know our sample mean is less than 100, but could that be by chance? How far would a mean from a sample of eight tend to stray from the population mean when chosen randomly from that population? Here's where we use the equation from earlier in this hack. We enter our sample values to produce our standard deviation of the mean, which is usually called the *standard error of the mean*:

$$\frac{\sigma}{\sqrt{n}} = \frac{4.53}{\sqrt{8}} = \frac{4.53}{2.83} = 1.60$$

We now know, thanks to the Central Limit Theorem, that most samples of eight Scouts will produce means that are within 1.6 IQ points of the population mean. It is unlikely, then, that our sample mean of 91.75 could have been drawn from a population with a mean of 100. A mean of 93, maybe, or 94, but not 100.

Because we know these means are normally distributed, we can use our knowledge of the shape of the normal distribution [Hack #23] to produce an exact probability that our mean of 91.75 could have come from a population with a mean of 100. It will happen way less than 1 out of 100,000 times. It seems very likely that my knot-tying students are tougher to teach than normal. I might ask for extra money.

Where Else It Works

A fuzzy version of the Central Limit Theorem points out that:

> Data that are affected by lots of random forces and unrelated events end up normally distributed.

As this is true of almost everything we measure, we can apply the normal distribution characteristics to make probability statements about most visible and invisible concepts.

We haven't even discussed the most powerful implication of the Central Limit Theorem. Means drawn randomly from a population will be normally distributed, *regardless of the shape of the population*. Think about that for a second. Even if the population from which you draw your sample of values is not normal—even if it is the opposite of normal (like my Uncle Frank, for example)—the means you draw out will still be normally distributed.

This is a pretty remarkable and handy characteristic of the universe. Whether I am trying to describe a population that is normal or non-normal, on Earth or on Mars, the trick still works.

HACK #3   Figure the Odds

Will I win the lottery? Will I get struck by lightning and hit by a bus on the same day? Will my basketball team have to meet our hated rival early in the NCAA tournament? At its core, statistics is all about determining the likelihood that something will happen and answering questions like these. The basic rules for calculating probability allow statisticians to predict the future.

This book is full of interesting problems that can be solved using cool statistical tricks. While all the tools presented in these hacks are applied in different ways in different contexts, many of the procedures used in these clever solutions work because of a common core set of elements: *the rules of probability*.

The *rules* are a key set of simple, established facts about how probability works and how probabilities should be calculated. Think of these two basic rules as a set of tools in a beginner's toolbox that, like a hammer and screwdriver, are probably enough to solve most problems:

Additive rule
   The probability of any one of several independent events occurring is the *sum* of each event's probability.

Multiplicative rule
   The probability of a series of independent events all occurring is the *product* of each event's probability.

These two tools will be enough to answer most of your everyday "What are the chances?" questions.

Questions About the Future

When a statistician says something like "a 1 out of 10 chance of happening," she has just made a prediction about the future. It might be a hypothetical statement about a series of events that will never be tested, or it might be an honest-to-goodness statement about what is about to happen. Either way, she's making a statistical statement about the likelihood of an outcome, which is just about all statisticians ever say [Hack #1].

If the following statement makes some intuitive sense to you, then you have all the ability necessary to act and think like a stat hacker: "If there are 10 things that might happen and all 10 things are equally likely to happen, then any 1 of those things has a 1 out of 10 chance of happening."

Research is full of questions that are answered using statistics, of course, and probability rules apply, but there are many problems in the world outside the laboratory that are more important than any stupid old science problem—like games with dice, for example! Imagine you are a part-time gambler, baby needs a new pair of shoes and all that, and the values showing the next time you throw a pair of dice will determine your future. You might want to know the likelihood of various outcomes of that dice roll. You might want to know that likelihood *very* precisely!

You can answer the three most important types of probability questions that you are likely to ask using only your two-piece probability toolkit. Your questions probably fall into one of these three types:

- How likely is it that a specific single outcome of interest will occur next? For example, will a dice roll of 7 come up next?

- How likely is it that any of a group of outcomes of interest will occur next? For example, will either a 7 or 11 come up next?

- How likely is it that a series of outcomes will occur? For example, could an honest pair of dice really be thrown all night and a 7 never (I mean *never!*) come up?! I mean, really, could it?! *Could it?!*

Probability Jargon

Before we talk about probability and how to determine it, we need to learn how to talk like a statistician. Remember the "1 out of 10 chance of happening" statement? Here are three ways of answering the question "What are the chances?":

As a percentage
 1 out of 10 can be expressed as 10 percent.

As odds
 The *odds* in a 1 out of 10 situation are 9 to 1 against—i.e., nine chances of losing against one chance of winning.

As a proportion
 10 percent can be expressed as 0.10. Technically, *probabilities* should be expressed as proportions or they should be called something else.

Likelihood of a Specific Outcome

When you are interested in whether something is likely to happen, that "something" can be called a *winning* event (if you are talking about a game) or just an *outcome of interest* (if you are talking about something other than a game). The primary principle in probability is that you divide the number of outcomes of interest by the total number of outcomes. The total number of outcomes is sometimes symbolized with an *S* (for set), and all the different outcomes of interest are sometimes symbolized as *A* (because it is the first letter of the alphabet, I guess; what am I, a mathematician?).

So, here's the basic equation for probability:

$$\frac{A}{S}$$

Figuring the chances of any particular outcome or event is a matter of counting the number of those outcomes, counting the number of all possible outcomes, and comparing the two. This is easily done in most situations with a small number of possible outcomes or a description of a winning outcome that is simple and involves a single event.

To answer a typical dice roll question, we can determine the chances of any specific value showing up on the next roll by counting the number of possible combinations of two six-sided dice that adds up to the value of interest. Then, divide that number by the total number of possible outcomes. With two 6-sided dice, there are 36 possible rolls.

For example, there are six ways to throw a 7 (I peeked ahead to Table 1-2), and 6/36 = .167, so the percentage chance of throwing a 7 on any single roll is about 17 percent.

 Calculate the total number of possible dice rolls, or outcomes, by multiplying the total number of sides on each die: $6 \times 6 = 36$.

Likelihood of a Group of Outcomes

If you are interested in whether any of a group of specific outcomes will occur, but you don't care which one, the additive rule states that you can figure your total probability by *adding together* all the individual probabilities. To answer our dice questions, Table 1-2 borrows some information from "Play with Dice and Get Lucky" [Hack #43] to express probability for various dice rolls as proportions.

Table 1-2. Probability of independent dice rolls

Dice roll	Number of outcomes	Probability
2	1	0.028
3	2	0.056
4	3	0.083
5	4	0.111

Table 1-2. Probability of independent dice rolls (continued)

Dice roll	Number of outcomes	Probability
6	5	0.139
7	6	0.167
8	5	0.139
9	4	0.111
10	3	0.083
11	2	0.056
12	1	0.028
Total	36	1.0

Table 1-2 provides information for various outcomes. For example, there are two different ways to roll a 3. Two winning outcomes divided by a total of 36 different possible outcomes results in a proportion of .056. So, about 6 percent of the time you'll roll a 3 with two dice. Notice also that the probabilities for every possible event add up to a perfect 1.0.

Let's apply the additive rule to see the chances of winning when, to win, we must get any one of several different dice rolls. If you will win with a roll of a 10, 11, *or* 12, for instance, add up the three individual probabilities:

.083 + .056 + .028 = .167

You will roll a 10, 11, or 12 about 17 percent of the time. The additive rule is used here because you are interested in whether any *one* of several *independent* events will happen.

Likelihood of a Series of Outcomes

What about when the probability question is whether *more* than one independent event will happen? This question is usually asked when you want to know whether a sequence of specific events will occur. The order of the events usually doesn't matter.

Using the data in Table 1-2 and the same three values of interest from our previous example (10, 11, and 12), we can figure the chance of a particular sequence of events occurring. What is the probability that, on a given series of three dice rolls in a row, you will roll a 10, an 11, *and* a 12? Under the multiplicative rule, multiply the three individual probabilities together:

.083 × .056 × .028 = .00013

This very specific outcome is very unlikely. It will happen less than .1 percent, or 1/10 of 1 percent of the time. The multiplicative rule is used here because you are interested in whether *all* of several *independent* events will happen.

What Probability Means

This hack talks about probability as the likelihood that something will happen. As I have placed our discussion within the context of analyzing possible outcomes, this is an appropriate way to think about probability. Among philosophers and social scientists who spend a lot of time thinking about concepts such as chance and the future and what's for lunch, there are two different views of probability.

Analytic view. This classic view of probability is the view of the mathematician and the approach used in this hack. The analytic view identifies all possible outcomes and produces a proportion of winning outcomes to all possible outcomes. That proportion is the probability.

We are predicting the future with the probability statement, and the accuracy of the prediction is unlikely to ever be tested. It is like when the weather forecaster says there is a 60 percent chance of rain. When it doesn't rain, we unfairly say the forecast was *wrong*, though, of course, we haven't really tested the accuracy of the probability statement.

Relative frequency view. Under the framework of this competing view, the probability of events is determined by collecting data and seeing what actually happened and how often it happened. If we rolled a pair of dice a thousand times and found that a 10 or an 11 or a 12 came up about 17 percent of the time, we would say that the chance of rolling one of those values is about 17 percent.

Our statement would really be about the past, not a prediction of the future. One might assume that past events give us a good idea of what the future holds, but who can know for sure? (Those of us who hold the analytic view of probability can know for sure, that's who.)

HACK #4 Reject the Null

Experimental scientists make progress by making a guess that they are sure is wrong.

Science is a goal-driven process, and the goal is to build a body of knowledge about the world. The body of knowledge is structured as a long list of scientific laws, rules, and theories about how things work and how they are. Experimental science introduces new laws and theories and tests them through a logical set of steps known as *hypothesis testing*.

Hypothesis Testing

A *hypothesis* is a guess about the world that is testable. For example, I might hypothesize that washing my car causes it to rain or that getting into a bathtub causes the phone to ring. In these hypotheses, I am suggesting a relationship between car washing and rainfall or between bathing and phone calls.

A reasonable way to see whether these hypotheses are true is to make observations of the variables in the hypothesis (for the sake of sounding like statisticians, we'll call that *collecting data*) and see whether a relationship is apparent. If the data suggests there is a relationship between my variables of interest, my hypothesis is supported, and I might reasonably continue to believe my guess is correct. If no relationship is apparent in the data, then I might wisely begin to doubt that my hypothesis is true or even reject it altogether.

There are four possible outcomes when scientists test hypotheses by collecting data. Table 1-3 shows the possible outcomes for this decision-making process.

Table 1-3. Possible outcomes of research hypothesis testing

	Hypothesis is correct: the world really is this way	Hypothesis is wrong: the world really is not this way
Data does support hypothesis: accept hypothesis	A. Correct decision: science makes progress.	B. Wrong decision: science is thwarted!
Data does not support hypothesis: reject hypothesis	C. Wrong decision: drat, foiled again!	D. Correct decision: science makes progress.

Outcomes A and D add to science's body of knowledge. Though A is more likely to make a research scientist all wriggly, D is just fine. Outcomes B and C, though, are mistakes, and represent misinformation that only confuses our understanding of the world.

Statistical Hypothesis Testing

The process of hypothesis testing probably makes sense to you—it is a fairly intuitive way to reach conclusions about the world and the people in it. People informally do this sort of hypothesis testing all the time to make sense of things.

Statisticians also test hypotheses, but hypotheses of a very specific variety. First, they have data that represents a sample of values from a real or theoretical population about which they wish to reach conclusions. So, their hypotheses are about populations. Second, they usually have hypotheses about the existence of a relationship among variables in the population of interest. A generic statistician's *research hypothesis* looks like this: there is a relationship between variable X and variable Y in the population of interest.

Unlike *research* hypothesis testing, with *statistical* hypothesis testing, the probability statement that a statistician makes at the end of the hypothesis testing process is not related to the likelihood that the research hypothesis is true. Statisticians produce probability statements about the likelihood that the research hypothesis is false. To be more technically accurate, statisticians make a statement about whether a hypothesis opposite to the research hypothesis is likely to be correct. This opposite hypothesis is typically a hypothesis of no relationship among variables, and is called the *null hypothesis*. A generic statistician's null hypothesis looks like this: there is no relationship between variable X and variable Y in the population of interest.

The research and null hypotheses cover all the bases. There either *is* or *is not* a relationship among variables. Essentially, when having to choose between these two hypotheses, concluding that one is false provides support for the other. Logically, then, this approach is just as sound as the more intuitive approach presented earlier and utilized naturally by humans every day. The preferred outcome by researchers conducting null hypothesis testing is a bit different than the general hypothesis-testing approach presented in Table 1-3.

As Table 1-4 shows, statisticians usually wish to reject their hypothesis. It is by rejecting the null that statistical researchers confirm their research hypotheses, get the grants, receive the Nobel prize, and one day are rewarded with their faces on a postage stamp.

Table 1-4. Possible outcomes of null hypothesis testing

	Null hypothesis is correct: there is no relationship in the population	Null hypothesis is wrong: there is a relationship in the population
Data does support null hypothesis: fail to reject the null	A. Correct decision: science makes progress.	B. Wrong decision: science is thwarted!
Data does not support null hypothesis: reject the null	C. Wrong decision: drat, foiled again!	D. Correct decision: science makes progress.

Although outcome A is still OK (as far as science is concerned), it is now outcome D that pleases researchers because it indicates support for their real guesses about the world, their research hypotheses. Outcomes B and C are still mistakes that hamper scientific progress.

Why It Works

Statisticians test the null hypothesis—guess the opposite of what they hope to find—for several reasons. First, proving something to be true is really, really tough, especially if the hypothesis involves a specific value, as statistical research often does. It is much easier to prove that a precise guess is wrong than prove that a precise guess is true. I can't prove that I am 29 years old, but it would be pretty easy to prove I am not.

It is also comparatively easy to show that any particular estimate of a population value is not likely to be correct. Most null hypotheses in statistics suggest that a population value is zero (i.e., there is no relationship between X and Y in the population of interest), and all it takes to reject the null is to argue that *whatever* the population value is, it probably isn't zero. Support for researchers' hypotheses generally come by simply demonstrating that the population value is greater than nothing, without specifically saying what that population value is exactly.

Quite a perk for the professional statistician, eh? All the statistician has to do is tell you that your answer is wrong, not tell you what the right answer is!

Even without using numbers as an example, philosophers of science have long argued that progress is best made in science by postulating hypotheses and then attempting to prove that they are wrong. For good science, *falsifiable* hypotheses are the best kind.

It is the custom to conduct statistical analyses this way: present a null hypothesis that is the opposite of the research hypothesis and see whether you can reject the null. R.A. Fisher, the early 20th century's greatest statistician, suggested this approach, and it has stuck. There are other methods, though. Plenty of modern statisticians have argued that we should concentrate on producing the best estimate of those population values of interest (such as the size of relationships among variables), instead of focusing on proving that the relationship is the size of some nonspecified number not equal to zero.

Go Big to Get Small

HACK #5

The best way to shrink your sampling error is to increase your sample size.

Whenever researchers are playing around with samples instead of whole populations, they are bound to make some mistakes. Because the basic trick of inferential statistics is to measure a sample and use the results to make guesses about a population [Hack #2], we know that there will always be some error in our guesses about the values in those populations. The good news is that we also know how to make the size of those errors as small as possible. The solution is to go big.

An early principle suggested in a gambling context was presented by Jakob Bernoulli (in 1713), who called his principle *the Golden Theorem*. It was later labeled by others (starting with Siméon-Denis Poisson in 1837) as the *Law of Large Numbers*. It is likely the single most useful discovery in the history of statistics and provides the basis for the key generic advice for all researchers: *increase your sample size!*

The early history of the science of applied statistics (we're talking the 17th and 18th centuries) is framed in the language of gambling and probability. This might be because it gave the gentlemen scholars of the time an excuse to combine their intellectual pursuits with pursuits of a less intellectual nature. The Laws of Probability, of course, are legitimately the mathematical basis for statistical procedures and inferences, so it might be that gambling applications were used simply as the best teaching examples for these central statistical concepts.

Laying Down the Law

One application of the Law is its effect on probability and occurrences. The Law includes the consequence that the increase in the accuracy of predicting outcomes governed by chance is a set amount. That is, the increase in accuracy is known. The expected distance between the probability of a certain outcome and the actual proportion of occurrences you observe decreases as the number of trials increases, and the exact size of this expected gap between expected and observed can be calculated. The generic name for this expected gap is the standard error **[Hack #18]**.

The size of the difference between the theoretical probability of an outcome and the proportion of times it actually occurs is proportional to:

$$\frac{1}{\sqrt{\text{Sample Size}}}$$

You can think of this formula as the mathematical expression of the Law of Large Numbers. For discussions of accuracy in the context of probability and outcome, the sample size is the number of trials. For discussion of accuracy in the context of sample means and population means, the sample size is the number of people (or random observations) in the sample.

Improving Accuracy

The specific values affected by the Law depend on the scale of measurement used and the amount of variability in a given sample. However, we can get a sense of the improvement or increase in accuracy made by various changes in sample sizes. Table 1-5 shows proportional increases in accuracy for all inferential statistics. So speaketh the Law.

Table 1-5. Effect of increasing sample size

Sample size	Relative decrease in error size	Meaning
1	1	The error is equal to the standard deviation of the variable in the population.
10	3.16	The error is about a third of its previous size. Just using 10 observations instead of 1 has dramatically increased our accuracy.
30	5.48	An increase from 1 to 30 people will dramatically improve accuracy. Even the jump from 10 to 30 is useful.
100	10	A sample of 100 people produces an estimate much closer to the population value (or expected probability). The size of the error with 100 people in a sample is just 1/10 of a standard deviation.
1,000	31.62	Estimates with so many observations are remarkably precise.

Why It Works

Let's look at this important statistical principle from several different angles. I'll state the law using three different approaches, beginning with the gambler's concerns, moving on to the issue of error, and ending with the implications for gathering a representative sample. All of the entries in this list are the exact same rule, just stated differently.

Gambling. If an event has a certain probability of occurring on a single trial, then the proportion of occurrences of the event over an infinite number of trials will equal that probability. As the number of trials approaches infinity, the proportion of occurrences approaches that probability.

Error. If a sample is infinitely large, the sample statistics will be equal to the population parameters. For example, the distance between the sample mean and the population mean decreases as the sample size approaches infinity. Errors in estimating population values shrink toward zero as the number of observations increases.

Implications. Samples are more representative of the population from which they are drawn when they include many people than when they include fewer people. The number of important characteristics in the population represented in a sample increases, as does the precision of their estimates, as the sample size gets larger.

> All these statements of the Law of Large Numbers are true only if based on the assumption that the occurrences or the sampling take place *randomly*.

In addition to providing the basis for calculations of standard errors, the Law of Large Numbers affects other core statistical issues such as power [Hack #8] and the likelihood of rejecting the null hypothesis when you should not [Hack #4]. Jakob Bernoulli's gambling pals might have been most interested in his Golden Theorem because they could get a sense of how many dice rolls it would take before the proportion of 7s rolled approached .166 or 16.6 percent, and could then do some solid financial planning.

For the last 300 years, though, all of social science has made use of this elegant tool to estimate how accurately something we see describes something we cannot see. Thanks, Jake!

See Also

- "Find Out Just How Wrong You Really Are" [Hack #18]

HACK
#6   Measure Precisely

Classical test theory provides a nice analysis of the components that combine to produce a score on any test. A useful implication of the theory is that the level of precision for test scores can be estimated and reported.

A good educational or psychological test produces scores that are *valid* and *reliable*. Validity is the extent to which the score on a test represents the level of whatever trait one wishes to measure, and the extent to which the test is useful for its intended purpose. To demonstrate validity, you must present evidence and theory to support that the interpretations of the test scores are correct.

Reliability is the extent to which a test consistently produces the same score upon repeated measures of the same person. Demonstrating reliability is a matter of collecting data that represent repeated measures and analyzing them statistically.

Classical Test Theory

Classical test theory, or *reliability theory*, examines the concept of a test score. Think of the observed score (the score you got) on a test you took sometime. Classical test theory defines that score as being made up of two parts and presents this theoretical equation:

Observed Score = True Score + Error Score

This equation is made up of the following elements:

Observed score
> The actual reported score you got on a test. This is typically equal to the number of items answered correctly or, more generally, the number of points earned on the test.

True score
> The score you should have gotten. This is not the score you *deserve*, though, or the score that would be the most valid. *True score* is defined as the average score you would get if you took the same test an infinite number of times. Notice this definition means that true scores represent only average performance and might or might not reflect the trait that the test is designed to measure. In other words, a test might produce true scores, but not produce *valid* scores.

Error Score
> The distance of your observed score from your true score.

Under this theory, it is assumed that performance on any test is subject to random error. You might guess and get a question correct on a social studies quiz when you don't really know the answer. In this case, the random error helps you.

> Notice this is still a measurement "error," even though it increased your score.

You might have cooked a bad egg for breakfast and, consequently, not even notice the last set of questions on an employment exam. Here, the random error hurt you. The errors are considered random, because they are not systematic, and they are unrelated to the trait that the test hopes to measure. The errors are considered errors because they change your score from your true score.

Over many testing times, these random errors should sometimes increase your score and sometimes decrease it, but across testing situations, the error should even out. Under classical test theory, reliability [Hack #31] is the extent to which test scores randomly fluctuate from occasion to occasion. A number representing reliability is often calculated by looking at the correlations among the items on the test. This index ranges from 0.0 to 1.0, with 1.0 representing a set of scores with no random error at all. The closer the index is to 1.0, the less the scores fluctuate randomly.

Standard Error of Measurement

Even though random errors should cancel each other out across testing situations, less than perfect reliability is a concern because, of course, decisions are almost always made based on scores from a single test administration. It doesn't do you any good to know that in the long run, your performance would reflect your true score if, for example, you just bombed your SAT test because the person next to you wore distracting cologne.

Measurement experts have developed a formula that computes a range of scores in which your true level of performance lies. The formula makes use of a value called the *standard error of measurement*. In a population of test scores, the standard error of measurement is the average distance of each person's observed score from that person's true score. It is estimated using information about the reliability of the test and the amount of variability in the group of observed scores as reflected by the standard deviation of those scores [Hack #2].

The formula for the standard error of measurement is:

$$\text{Standard Error} = \text{Standard Deviation}\sqrt{1 - \text{Reliability}}$$

Here is an example of how to use this formula. The Graduate Record Exam (GRE) tests provide scores required by many graduate schools to help in making admission decisions. Scores on the GRE Verbal Reasoning test range from 200 to 800, with a mean of about 500 (it's actually a little less than that in recent years) and a standard deviation of 100.

Reliability estimates for scores from this test are typically around .92, which indicates very high reliability. If you receive a score of 520 when you take this exam, congratulations, you performed higher than average. 520 was your observed score, though, and your performance was subject to random error. How close is 520 to your true score? Using the standard error of measurement formula, our calculations look like this:

1. $1 - .92 = .08$
2. The square root of .08 is .28
3. $100 \times .28 = 28$

The standard error of measurement for the GRE is about 28 points, so your score of 520 is most likely within 28 points of what you would score on average if you took the test many times.

Building Confidence Intervals

What does it mean to say that an observed score is most likely within one standard error of measurement of the true score? It is accepted by measurement statisticians that 68 percent of the time, an observed score will be within one standard error of measurement of the true score. Applied statisticians like to be more than 68 percent sure, however, and usually prefer to report a range of scores around the observed score that will contain the true score 95 percent of the time.

To be 95 percent sure that one is reporting a range of scores that contain an individual's true score, one should report a range constructed by adding and subtracting about *two* standard errors of measurement. Figure 1-1 shows what confidence intervals around a score of 520 on the GRE Verbal test look like.

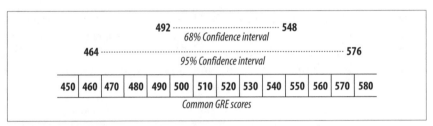

Figure 1-1. Confidence intervals for a GRE score of 520

Why It Works

The procedure for building confidence intervals using the standard error of measurement is based on the assumptions that errors (or error scores) are random, and that these random errors are normally distributed. The normal curve [Hack #25] shows up here as it does all over the world of human characteristics, and its shape is well known and precisely defined. This precision allows for the calculation of precise confidence intervals.

The standard error of measurement is a standard deviation. In this case, it is the standard deviation of error scores around the true score. Under the normal curve, 68 percent of values are within one standard deviation of the mean, and 95 percent of scores are within about two standard deviations (more exactly, 1.96 standard deviations). It is this known set of probabilities that allows measurement folks to talk about 95 percent or 68 percent confidence.

What It Means

How is knowing the 95 percent confidence interval for a test score helpful? If you are the person who is requiring the test and using it to make a decision, you can judge whether the test taker is likely to be within reach of the level of performance you have set as your standard of success.

If you are the person who took the test, then you can be pretty sure that your true score is within a certain range. This might encourage you to take the test again with some reasonable expectation of how much better you are likely to do by chance alone. With your score of 520 on the GRE, you can be 95 percent sure that if you take the test again right away, your new score could be as high as 576. Of course, it could drop and be as low as 464 the next time, too.

H A C K   Measure Up
#7
Four levels of measurement determine how the scores produced in measurement can be used. If you have not measured at the right level, you might not be able to play with those scores the way you want.

Statistical procedures analyze numbers. The numbers must have meaning, of course; otherwise, the exercises are of little value. Statisticians call numbers with meaning *scores*. Not all the scores used in statistics, however, are created equal. Scores have different amounts of information in them, depending on the rules followed for creating the scores.

When you decide to measure something, you must choose the rules by which you assign scores very carefully. The *level of measurement* determines which sorts of statistical analyses are appropriate, which will work, and which will be meaningful.

> *Measurement* is the meaningful assignment of numbers to things. The things can be concrete objects, such as rocks, or abstract concepts, such as intelligence.

Here's an example of what I mean when I say not all scores are created equal. Imagine your five children took a spelling test. Chuck scored a 90, Dick and Jan got 80s, Bob scored 75, and Don got only 50 out of 100 correct. If a friend asked how your kids did on the big test, you might report that they averaged 75. This is a reasonable summary. Now, imagine that your five children ran a foot race against each other. Bob was first, Jan second, Dick third, Chuck fourth, and Don fifth. Your nosey friend again asks how they did. With a proud smile, you report that they averaged third place. This is not such a reasonable summary, because it provides no information.

In both cases, though, scores were used to indicate performance. The difference lies only in the level of measurement used.

There are four levels of measurement—that is, four ways that numbers are used as scores. The levels differ in the amount of information provided and the types of mathematical and statistical analyses that can be meaningfully conducted on them. The four levels of measurement are *nominal*, *ordinal*, *interval*, and *ratio*.

Using Numbers as Labels

If you are planning to use scores to indicate only that the things belong to different groups, measure at the *nominal* level. The nominal level of measurement uses numbers only as *names*: labels for various categories (*nominal* means "in name only").

For example, a scientist who collects data on men and women, using a 1 to indicate a male subject and a 2 to indicate a female subject, is using the numbers at a nominal level. Notice that even though the number 2 is mathematically greater than the number 1, a 2 in this data set does not mean *more* of anything. It is used only as a name.

Using Numbers to Show Sequence

If you want to analyze your scores in ways that rely on performance measured as evidence of some sequence or order, measure at the *ordinal* level. Ordinal measurement provides all the information the nominal level provides, but it adds information about the order of the scores. Numbers with greater values can be compared with numbers at lower values, and the people or otters or whatever was measured can be placed into a meaningful order.

Take, for example, your rank order in your high school class. The valedictorian is usually the person who received a score of 1 when grade point averages are compared. Notice that you can compare scores to each other, but you don't know anything about the distance between the scores. In a footrace, the first-place finisher might have been just a second ahead of the second-place runner, while the second-place runner might have been 30 seconds ahead of the runner who came in third place.

Using Numbers to Show Distance

Interval level measurement uses numbers in a way that provides all the information of earlier levels, but adds an element of precision. This level of measurement produces scores that are interpreted as having an equal difference between any two adjacent scores.

For example, on a Fahrenheit thermometer, the meaningful difference between 70 and 69 degrees—1 degree—is equal to the difference between 32 and 31 degrees. That one degree is assumed to be the same amount of heat (or, if you prefer, pressure on the liquid in the thermometer), regardless of where on the scale the interval exists.

The interval level provides much more information than the ordinal level, and you can now meaningfully average scores. Most educational and psychological measurement takes place at the interval level.

Though interval level measurement would seem to solve all of our problems in terms of what we can and cannot do statistically, there are still some mathematical operations that are not meaningful at this level. For instance, we don't make comparisons using fractions or proportions. Think about the way we talk about temperature. If a 40-degree day follows an 80-degree day, we do not say, "It is half as hot today as yesterday." We also don't refer to a student with a 120 IQ as "one-third smarter" than a student with a 90 IQ.

The word *interval* is a term from old-time castle architecture. You know those tall towers or turrets where archers were stationed for defense? Around the circular tops, there was typically a pattern of a protective stone, then a gap for launching arrows, followed by another protective stone, and so on. The gaps were called intervals ("between walls"), and the best designed defenses had the stones and gaps at equal intervals to provide 360-degree protection.

Using Numbers to Count in Concrete Ways

The highest level of measurement, *ratio*, provides all the information of the lower levels but also allows for proportional comparisons and the creation of percentages. Ratio level measurement is actually the most common and intuitive way in which we observe and take accounting of the natural world. When we count, we are at the ratio level. How many dogs are on your neighbor's porch? The answer is at the ratio level.

Ratio level measurement provides so much information and allows for all possible statistical manipulations because ratio scales use a *true zero*. A true zero means that a person could score 0 on the scale and really have zero of the characteristic being measured. Though a Fahrenheit temperature scale, for example, does have a zero on it, a zero-degree day does not mean there is absolutely no heat. On interval scales, such as in our thermometer example, scores can be negative numbers. At the ratio level of measurement, there are no negative numbers.

Choosing Your Level of Measurement

Which level of measurement is right for you? Because of the advantages of moving to at least the interval level, most social scientists prefer to measure at the interval or ratio level. At the interval level, you can safely produce descriptive statistics and conduct inferential statistical analyses, such as *t* tests, analyses of variance, and correlational analyses. Table 1-6 provides a summary of the strengths and weaknesses of each level of measurement.

Table 1-6. Levels of measurement

Level of measurement	Strength	Weakness
Nominal	Describes categorical data.	Numbers do not indicate quantity.
Ordinal	Allows comparison between scores.	Difficult to summarize scores.
Interval	Most statistical analyses are possible.	Proportional comparisons are not possible.
Ratio	True zero allows for all possible statistical analyses.	Some variables of interest do not have a true zero.

To choose the correct statistical analysis of data created by others, identify the level of measurement used and benefit from its strengths. If you are creating the data yourself, consider *measuring up*: using the highest level of measurement that you can.

Controversial Tools

Since the levels of measurement became commonly accepted in the 1950s, there has been some debate about whether we really need to clearly be at the interval level to conduct statistical analyses. There are many common forms of measurement (e.g., attitude scales, knowledge tests, or personality measures) that are not unequivocally at the interval level, but might be somewhere near the top of the ordinal level range. Can we safely use this level of data in analyses requiring interval scaling?

A majority consensus in the research literature is that if you are at least at the ordinal level and believe that you can make meaning out of interval-level statistical analyses, then you can safely perform inferential statistical analyses on this type of data. In the real world of research, by the way, almost everybody chooses this approach (whether they know it or not).

The basic value of making analytical decisions based on level of measurement is hard to deny, however. A classic example of the importance of measurement levels is described by Frederick Lord in his 1953 article "On the Statistical Treatment of Football Numbers" (*American Psychologist*, Vol. 8,

750–751). An absent-minded statistician eagerly analyzes some data given him concerning the college football team, and produces a report full of means and standard deviations and other sophisticated analyses. The data, though, turn out to be the numbers from the backs of the players' jerseys. A clear instance of not paying attention to level of measurement, perhaps, but the statistician stands by his report. The numbers themselves, he explains, don't know where they came from; they behave the same way regardless.

H A C K #8   Power Up

Success in social science research is typically defined by the discovery of a statistically significant finding. To increase the chances of finding something, anything, the primary goal of the statistically savvy super-scientist should be to increase power.

There are two potential pitfalls when conducting statistically based research. Scientists might decide that they have found something in a population when it really exists only in their sample. Conversely, scientists might find nothing in their sample when, in reality, there was a beautiful relationship in the population just waiting to be found.

The first problem is minimized by sampling in a way that represents the population [Hack #19]. The second problem is solved by increasing *power*.

Power

In social science research, a statistical analysis frequently determines whether a certain value observed in a sample is likely to have occurred by chance. This process is called a *test of significance*. Tests of significance produce a *p-value* (probability value), which is the probability that the sample value could have been drawn from a particular population of interest.

The lower the p-value, the more confident we are in our beliefs that we have achieved statistical significance and that our data reveals a relationship that exists not only in our sample but also in the whole population represented by that sample. Usually, a predetermined level of significance is chosen as a standard for what counts. If the eventual p-value is equal to or lower than that predetermined level of significance, then the researcher has achieved a level of significance.

Statistical analyses and tests of significance are not limited to identifying relationships among variables, but the most common analyses (*t* tests, *F* tests, chi-squares, correlation coefficients, regression equations, etc.) usually serve this purpose. I talk about relationships here because they are the typical *effect* you're looking for.

The *power* of a statistical test is the probability that, given that there *is* a relationship among variables in the population, the statistical analysis will result in the decision that a level of significance has been achieved. Notice this is a conditional probability. There must be a relationship in the population to find; otherwise, power has no meaning.

Power is not the chance of finding a significant result; it is the chance of finding that relationship *if* it is there to find. The formula for power contains three components:

- Sample size
- The predetermined level of significance (p-value) to beat (be less than)
- The *effect size* (the size of the relationship in the population)

Conducting a Power Analysis

Let's say we want to compare two different sample groups and see whether they are different enough that there is likely a real difference in the populations they represent. For example, suppose you want to know whether men or women sleep more.

The design is fairly straightforward. Create two samples of people: one group of men and one group of women. Then, survey both groups and ask them the typical number of hours of sleep they get each night. To find any real differences, though, how many people do you need to survey? This is a power question.

A *t* test compares the mean performance of two sample groups of scores to see whether there is a significant difference [Hack #17]. In this case, statistical significance means that the difference between scores in the two populations represented by the two sample groups is probably greater than zero.

Before a study begins, a researcher can determine the power of the statistical analysis that will be used. Two of the three pieces needed to calculate power are already known before the study begins: you can decide the sample size and choose the predetermined level of significance. What you can't know is the true size of the relationship between the variables, because data for the planned research has not yet been generated.

The size of the relationship among the variables of interest (i.e., the effect size) can be estimated by the researcher before the study begins; power also can be estimated before the study begins. Usually, the researcher decides on the smallest relationship size that would be considered important or interesting to find.

Once these three pieces (sample size, level of significance, and effect size) are determined, the fourth piece (power) can be calculated. In fact, setting the level of any three of these four pieces allows for calculation of the fourth piece. For example, a researcher often knows the power she would like an analysis to have, the effect size she wants to be declared statistically significant, and the preset level of significance she will choose. With this information, the researcher can calculate the necessary sample size.

> For estimating power, researchers often use a standard accepted procedure that identifies a power goal of .80 and assigns a preset level of significance of .05. A power of .80 means that a researcher will find a relationship or effect in her sample 80 percent of the time *if* there is such a relationship in the population from which the sample was drawn.

The effect size (or index of relationship size [Hack #10]) with *t* tests is often expressed as the difference between the two means divided by the standard deviation in each group. This produces effect sizes in which .2 is considered small, .5 is considered medium, and .8 is considered large. The power analysis question is: how big a sample in each of the two groups (how many people) do I need in order to find a significant difference in test scores?

The actual formula for computing power is complex, and I won't present it here. In real life, computer software or a series of dense tables in the back of statistics books are used to estimate power. I have done the calculations for a series of options, though, and present them in Table 1-7. Notice that the key variables are effect size and sample size. By convention, I have kept power at .80 and level of significance at .05.

Table 1-7. Necessary sample sizes for various effect sizes

Effect size	Sample size
.10	1,600
.20	400
.30	175
.40	100
.50	65
1.0	20

Imagine that you think the actual difference in your gender-and-sleep study will be real, but small. A difference of about .2 standard deviations between groups in *t* test analyses is considered small, so you might expect a .2 effect size. To find that small of an effect size, you need 400 people in *each* group! As the effect size increases, the necessary sample size gets smaller. If the population effect size is 1.0 (a very large effect size and a big difference between the two groups), 20 people per group would suffice.

Making Inferences About Beautiful Relationships

Scientists often rely on the use of statistical inference to reject or accept their research hypotheses. They usually suggest a null hypothesis that says there is no relationship among variables or differences between groups. If their sample data suggests that there is, in fact, a relationship between their variables in the population, they will reject the null hypothesis [Hack #4] and accept the alternative, their research hypothesis, as the best guess about reality.

Of course, mistakes can be made in this process. Table 1-8 identifies the possible types of errors that can be made in this hypothesis-testing game. Rejecting the null hypothesis when you should not is called a *Type I error* by statistical philosophers. Failing to reject the null when you should is called a *Type II error*.

Table 1-8. Errors in hypothesis testing

Action	Null hypothesis is true	Null hypothesis is false
Reject null hypothesis	Type I error	Significant finding
Fail to reject null	Correct decision	Type II error

What you want to do as a smart scientist is avoid the two types of errors and produce a significant finding. Reaching a correct decision to not reject the null when the null is true is okay too, but not nearly as fun as a significant finding. "Spend your life in the upper-right quadrant of the table," my Uncle Frank used to say, "and you will be happy and wealthy beyond your wildest dreams!"

To have a good chance of reaching a statistically significant finding, one condition beyond your control must be true. The null hypothesis must be false, or your chances of "finding" something are slim. And, if you do "find" something, it's not really there, and you will be making a big error—a Type I error. There must actually be a relationship among your research variables in the population for you to find it in your sample data.

So, fate decides whether you wind up in the column on the right in Table 1-8. *Power* is the chance of moving to the top of that column once you get there. In other words, power is the chance of correctly rejecting the null hypothesis when the null hypothesis is false.

Why It Works

This relationship between effect size and sample size makes sense. Think of an animal hiding in a haystack. (The animal is the *effect size*; just work with me on this metaphor, please.) It takes fewer observations (handfuls of hay) to find a big ol' effect size (like an elephant, say) than it would to find a tiny animal (like a cute baby otter, for instance). The number of people represents the number of observations, and big effect sizes hiding in populations are easier to find than smaller effect sizes.

The general relationship between effect size and sample size in power works the other way, too. Guess at your effect size, and just increase your sample size until you have the power you need. Remember, Table 1-7 assumes you want to have 80 percent power. You can always work with fewer people; you'll just have less power.

Where It Doesn't Work

It is important to remember that power is not the chance of success. It is not even the chance that a level of significance will be reached. It is the chance that a level of significance will be reached if all the values estimated by the researcher turn out to be correct. The hardest component of the formula to guess or set is the effect size in the population. A researcher seldom knows how big the thing is that he is looking for. After all, if he did know the size of the relationship between his research variables, there wouldn't be much reason to conduct the study, would there?

HACK   # Show Cause and Effect

#9   Statistical researchers have established some ground rules that must be followed if you hope to demonstrate that one thing causes another.

Social science research that uses statistics operates under a couple of broad goals. One goal is to collect and analyze data about the world that will support or reject hypotheses about the relationships among variables. The second goal is to test hypotheses about whether there are cause-and-effect relationships among variables. The first goal is a breeze compared to the second.

There are all sorts of relationships between things in the world, and statisticians have developed all sorts of tools for finding them, but the presence of a relationship doesn't mean that a particular variable causes another. Among humans, there is a pretty good positive correlation [Hack #11] between height and weight, for example, but if I lose a few pounds, I won't get shorter. On the other hand, if I grow a few inches, I probably will gain some weight.

Knowing only the correlation between the two, however, can't really tell me anything about whether one thing caused the other. Then again, the *absence* of a relationship would seem to tell me about cause and effect. If there is no correlation between two variables, that would seem to rule out the possibility that one causes the other. The presence of the correlation allows for that possibility, but does not prove it.

Designing Effective Experiments

Researchers have developed frameworks for talking about different research designs and whether such designs even allow for proof that one variable affects another. The different designs involve the presence or absence of comparison groups and how participants are assigned to those groups.

There are four basic categories of group designs, based on whether the design can provide strong evidence, moderate evidence, weak evidence, or no evidence of cause and effect:

Non-experimental designs
> These designs usually involve just one group of people, and statistics are used to either describe the population or demonstrate a relationship between variables. An example of this design is a correlational study, where simple associations among variables are analyzed [Hack #11]. This type of design provides no evidence of cause and effect.

Pre-experimental designs
> These designs usually involve one group of people and two or more measurement occasions to see whether change has occurred. An example of this design is to give a *pretest* to a group of people, do something to them, give them a *post-test*, and see whether their scores change. This type of design provides weak evidence of cause and effect because forces other than whatever you did to the poor folks could have caused any change in scores.

Quasi-experimental designs
> These designs involve more than one group of people, with at least one group acting as a comparison group. Assignment to these groups is not random but is determined by something outside the researcher's control. An example of this design is comparing males and females on their attitudes toward statistics. At best, this sort of design provides moderate evidence of cause and effect. Without random assignment to groups, the groups are likely not equal on a bunch of unmeasured variables, and those might be the real cause for any differences that are found.

Experimental designs
> These designs have a comparison group and, importantly, people are assigned to the groups randomly. The random assignment to groups allows for researchers to assume that all groups are equal on all unmeasured variables, thus (theoretically) ruling them out as alternative

explanations for any differences found. An example of this design is a drug study in which all participants randomly get either the drug being tested or a comparison drug or a placebo (sugar pill).

Does Weight Cause Height?

Earlier in this hack, I mentioned a well-known correlational finding: in people, height and weight tend to be related. Taller males weigh more, usually, then shorter males, for example. I laughed off the suggestion that if we fed people more, they would get taller—because of what I think I know about how the body grows, the suggestion that weight causes height is theoretically unlikely. But what if you demanded *scientific* proof?

I could test the hypothesis that weight causes height using a basic *experimental* design. Experimental designs have a comparison group, and the assignment to such groups must be random. Any relationships found under such circumstances are likely causal relationships. For my study, I'd create two groups:

Group 1
> Thirty college freshmen, who I would recruit from the population of the Midwestern university where I work. This group would be the *experimental* group; I would increase their weight and measure whether their height increases.

Group 2
> Thirty college freshmen, who I would recruit from the population of the Midwestern university where I work. This group would be the *control* group; I would not manipulate their weight at all and would then measure whether their height changes.

> In this design, scientists would call weight the *independent variable* (because we don't care what causes *it*) and height the *dependent variable* (because we wonder whether it *depends* on, or is caused by, the independent variable).

Because this design matches the criteria for experimental designs, we could interpret any relationships found as evidence of cause and effect.

Fighting Threats to Validity

Research conclusions fall into two types. They have to do with the cause-and-effect claim and whether any such claim, once it is established, is generalizable to whole populations or outside the laboratory. Table 1-9 displays the primary types of validity concerns when interpreting research results. These concerns are the hurdles that must be crossed by researchers.

Table 1-9. Validity of research results

Validity concern	Validity question
Statistical conclusion validity	Is there a relationship among variables?
Internal validity	Is the relationship a cause-and-effect relationship?
Construct validity	Is the cause-and-effect relationship among the variables you believe should be affected?
External validity	Does this cause-and-effect relationship exist everywhere for everyone?

Even when researchers have chosen a true experimental design, they still must worry that any results might not really be due to one variable *affecting* another. A cause-and-effect conclusion has many threats to its validity, but fortunately, just by thinking about it, researchers have identified many of these threats and have developed solutions.

> Researchers' understanding of group designs, the terminology used to describe them, the identification of threats to validity in research design, and the tools to guard against the threats are pretty much entirely due to the extremely influential works of Cook and Campbell, cited in the "See Also" section of this hack.

A few threats to the validity of causal claims and claims of generalizability are discussed next, along with some ways of eliminating them. There are dozens of threats identified and dealt with in the research design literature, but most of them are either unsolvable or can be solved with the same tools described here:

History

Outside events could affect results. A solution is to use a control group (a comparison group that does not receive the drug or intervention or whatever), with random assignment of subjects to groups. Another part of the solution is to control both groups' environments as much as possible (e.g., in laboratory-type settings).

Maturation

Subjects develop naturally during a study, and changes might be due to these natural developments. Random assignment of participants to an experimental group and a control group solves this problem nicely.

Selection

There might be systematic bias in assigning subjects to groups. The solution is to assign subjects randomly.

Testing

Just taking a pretest might affect the level of the research variable. Create a comparison group and give both groups the pretest, so any changes will be equal between the groups. And assign subjects to the two groups randomly (are you starting to see a pattern here?).

Instrumentation

There might be systematic bias in the measurement. The solution is to use valid, standardized, objectively scored tests.

Hawthorne Effect

Subjects' awareness that they are subjects in a study might affect results. To fight this, you could limit your subjects' awareness of what results you expect, or you could conduct a double-blind study in which subjects (and researchers) don't even know what treatment they are receiving.

The validity of research design and the validity of any claims about cause and effect are similar to claims of validity in measurement **[Hack #28]**. Such arguments are open and unending, and validity conclusions rest on a reasoned examination of the evidence at hand and consideration for what seems reasonable.

See Also

- Campbell, D.T. and Stanley, J.C. (1966). *Experimental and quasi-experimental designs for research.* Chicago: Rand McNally.
- Cook, T.D. and Campbell, D.T. (1979). *Quasi-experimentation: Design and analysis issues for field settings.* Boston: Houghton-Mifflin.
- Shadish, W.R., Cook, T.D., and Campbell, D.T. (2002). *Experimental and quasi-experimental designs for generalized causal inference.* Boston: Houghton-Mifflin.

HACK #10   Know Big When You See It

You've just read about an amazing new scientific discovery, but is such a finding really a big deal? By applying effect size interpretations, you can judge the importance of such announcements (or lack thereof) for yourself.

Something is missing in most reports of scientific findings in nonscientific publications, on TV, on the radio, and—do I even have to mention—on the Web. Although reports in such media typically do a pretty good job of only reporting findings that are "statistically significant," this is not enough to determine whether anything really important or useful has been discovered. A big drug study can report "significant" results, but still not have found anything of interest to the rest of us or even other researchers.

As we repeat in many places in this book, significance [Hack #4] means only that what you found is likely to be true about the bigger population you sampled from. The problem is that this fact alone is not nearly enough for you to know whether you should change your behavior, start a new diet, switch drugs, or reinterpret your view of the world.

What you need to know to make decisions about your life and reality in light of any new scientific report is the *size* of the relationship that has just been brought to light. How *much* better is brand A than brand B? How *big* is that SAT difference between boys and girls in meaningful terms? Is it worth it to take that half an aspirin a day, every day, to lower your risk of a heart attack? How much lower *is* that risk anyway?

The strength of that relationship should be expressed in some standardized way, too. Otherwise, there is no way to really judge how big it is. Using a statistical tool known as *effect size* will let you know big when you see it.

Seeing Effect Sizes Everywhere

An effect size is a standardized value that indicates the strength of a relationship between two variables. Before we talk about how to recognize or interpret effect sizes, let's begin with some basics about relationships and statistical research.

Statistical research has always been interested in relationships among variables. The correlation coefficient, for example, is an index of the strength and direction of relationships between two sets of scores [Hack #11]. Less obvious, but still valid, examples of statistical procedures that measure relationships include *t* tests [Hack #17] and analysis of variance, a procedure for comparing more than two groups at one time.

 Even procedures that compare different groups are still interested in relationships between variables. With a *t* test, for instance, a significant result means that it matters which group a person is in. In other words, there is an association between the independent variable (which defines the groups) and the dependent variable (the measured outcome).

Finding or Computing Effect Sizes

This hack is about finding and interpreting effect sizes to judge the implications of scientific findings reported in the popular media or in scientific writings. Often, the effect size is reported directly and you just have to know how to interpret it. Other times, it is not reported, but enough information is provided so that you can figure out what the effect size is.

When effect sizes are reported, they are typically one of three types. They differ depending on the procedure used and the way that procedure quantifies the information of interest. In each case, the effect size can be interpreted as estimates of the "size of the relationship between variables." Here are the three typical types of effect sizes:

Correlation coefficient
A *correlation*, symbolized by r, is already a measure of the relationship between variables and, thus, *is* an effect size. Because correlations can be negative, though, the value is sometimes squared to produce a value that is always greater than zero. Thus, the value of r^2 is interpreted as the "proportion of variance" shared by variables.

d

This value, symbolized by d strangely enough, summarizes the difference between two group means used in a t test. It is calculated by dividing the mean difference of the two groups by the average standard deviation in the two groups.

> Here's an alternative, easy, super-fun, ultra-cool, and neato-swell way to calculate d:
>
> $$d = t \sqrt{\frac{\text{Sample Size in Group 1} + \text{Sample Size in Group 2}}{(\text{Sample Size in Group 1})(\text{Sample Size in Group 2})}}$$

Eta-squared
The effect size most often reported for the results of an analysis of variance is symbolized as η^2. Similar to r^2, it is interpreted as the "proportion of variance" in the dependent variable (the outcome variable) accounted for by the independent variable (what group you are in).

Interpreting Effect Sizes

With levels of significance, statisticians have adopted certain sizes that are "good" to achieve. For example, most statistical researchers hope to achieve a .05 or lower level of significance. With effect sizes, though, there are not always certain values that are clearly good or clearly bad. Still, some standards for small, medium, and large effect sizes have been suggested.

The standards for big, medium, and little are based, for the most part, on the effect sizes that are normally found in real-world research. If a given effect size is so large as to be rarely found in published research, it is considered to be big. If the effect size is tiny and easy to find in real-life research, then it is considered to be small.

You should decide yourself, though, how big an effect size is of interest to you when interpreting research results. It all depends on the area of investigation. Table 1-10 provides the rules of thumb for how big is big.

Table 1-10. Effect size standards

Effect sze	Small	Medium	Large
r	+/-.10	+/-.30	+/-.50
r^2	.01	.09	.25
d	.2	.5	.8
η^2	.01	.06	.14

Interpreting Research Findings

The advantage of talking about effect sizes when discussing research results is that everyone can get a sense of what impact the given research variable (or intervention, or drug, or teaching technique) is really having on the world. Because they are typically reported without any probability information (level of significance), effect sizes are most useful when provided alongside traditional level of significance numbers. This way, two questions can be answered:

- Does this relationship probably exist in the population?
- How big is the relationship?

Remember our example of whether you should decide to take half an aspirin each day to cut down your chances of having a heart attack? A well-publicized study in the late 1980s found a statistically significant relationship between these two variables. Of course, you should talk with your doctor before you make any sort of decision like this, but you should also have as much information as possible to help you make that decision. Let's use effect size information to help us interpret these sorts of findings.

Here is what was reported in the media:

A sample of 22,071 physicians were randomly divided into two groups. For a long period of time, half took aspirin every day, while the other half took a placebo (which looked and tasted just like aspirin). At the end of the study period (which actually ended early because the effectiveness of aspirin was considered so large), the physicians taking aspirin were about half as likely to have had a heart attack than the placebo group. 1.71 percent of the placebo physicians had attacks versus about 1 percent (.94 percent) of the aspirin physicians. The findings were statistically significant.

The "clear" interpretation of such findings is that taking aspirin cuts your chances of a heart attack in half. Assuming that the study was representative and the physicians in the study are like you and me in important ways, this interpretation is fairly correct.

Another way to interpret the findings is to look at the effect size of the aspirin use. Using a formula for proportional comparisons, the effect size for this study is .06 standard deviations, or a d of .06. Applying the effect size standards shown in Table 1-10, this effect size should be interpreted as small— very small, really. This interpretation suggests that there is really quite a tiny relationship between aspirin-taking and heart attacks. The relationship is real, just not very strong.

One way to think about this is that your chances of having a heart attack during a given period of time is pretty small to begin with. 98.76 percent of everyone in the study did not have a heart attack, whether they took aspirin or not. Although taking aspirin does lower your chances, they go from small to a little smaller. It is similar to the idea that entering the lottery massively increases your chances of winning compared to those who do not enter, but your chances are still slim.

Why It Works

A researcher can achieve significant results, but still not have found anything for anyone to get excited about. This is because *significance* tells you only that your sample results probably did not occur by chance. The results are real and likely exist in the population. If you have found evidence of a *small* relationship between two variables or between the use of a drug and some medical outcome, the relationship might be so small that no one is really interested in it. The effect of the drug might be real, but weak, so it's not worth recommending to patients. The relationship between A and B might be greater than zero, but so tiny as to do little to help understand either variable.

Modern researchers are still interested in whether there is statistical significance in their findings, but they should almost always report and discuss the effect size. If the effect size is reported, you can interpret it. If it is not reported, you can often dig out the information you need from published reports of scientific findings and calculate it yourself. The cool part is that you might then know more about the importance of the discovery than the media who reported the findings and, maybe, even the scientists themselves.

Discovering Relationships
Hacks 11–22

There are invisible webs of relationships around us. Variable A causes Variable B, which influences Variable C, which is entirely independent of Variable D, unless Variable E comes into play. The hacks in this chapter allow you to discover these connections and describe them accurately. These are the hacks that reveal the hidden reasons for why people do the things they do and why things are the way they are.

The connections between one trait and another, between a cause and an effect, are relationships that are easily revealed—with the right tricks. Begin by identifying the strength of any association [Hack #11], and then draw what it looks like [Hack #12]. Next, use your knowledge of that relationship to make predictions [Hack #13], and then improve the accuracy of those predictions [Hack #14]. Some relationships appear through the observation of unexpected occurrences [Hacks #15 and #16] or by noticing real differences between groups [Hack #17].

Because we cannot measure every example of a person, fish, or pine tree that we might be interested in, we must rely on representative samples [Hack #19] to provide our observations. Sampling can mislead us [Hack #18], however, or it can work in surprisingly cool ways [Hack #20].

To share your findings with others or understand what these findings have to tell you, you need to avoid both being deceived and deceiving others. Be careful not to misinterpret any numbers [Hack #21] or pictures [Hack #22].

Pack these tools in your tool belt and head out to find whatever there is to find.

Discover Relationships

Revealing the invisible connections in the world is just a matter of recording
observations and computing the magical, mystical correlation coefficient.

You probably make all sorts of assumptions about why people feel the way
they feel or do the things they do. Statistical researchers would call these
assumptions *hypotheses* about the relationship among *variables*.

Regardless of what science calls it, you probably do it. You might make
these guesses about associations between attitudes and behavior or between
attitudes and attitudes or behaviors and behaviors. You might do it informally as you seek to understand people in the world around you, or you
might need to do it as a marketing specialist to understand your customer,
or you might be a struggling psychology graduate student who needs to
complete a class assignment that requires statistical analysis of the relationship between self-esteem and depression.

In statistics, such a relationship is called a *correlation*. The number describing the size of that relationship is a *correlation coefficient*. By computing this
useful value, you can get answers to any question you have about relationships (except in terms of dating relationships; you're on your own there).

Testing Hypotheses About Relationships

Imagine a study in which a researcher for the American Cheesecake Sellers
Association has a hypothesis that the reason people like cheesecake is that
they like cheese. She is guessing that there is a relationship between attitude
toward cheese and attitude toward cheesecake. If her hypothesis turns out to
be correct, she'll purchase the huge mailing list of cheese lovers from the
American Cheese Lovers Association and send them informative brochures
about the healing properties of cheesecake. If she's right, sales will rocket up!

To test her hypothesis, she creates two surveys. One asks respondents to say
how they feel about cheese, and the other asks how they feel about cheesecake. A score of 50 means the person loves cheese (or cheesecake), and a
score of 0 means the person hates cheesecake (or cheese). Table 2-1 shows
the results for the data she collects from five people on the bus on her way to
work.

Table 2-1. Data for the relationship between cheese and cheesecake attitudes

Person	Attitude toward cheese	Attitude toward cheesecake
Larry	50	36
Moe	45	35

Table 2-1. Data for the relationship between cheese and cheesecake attitudes (continued)

Person	Attitude toward cheese	Attitude toward cheesecake
Curly Joe	30	22
Shemp	30	25
Groucho	10	20

Let's look at the data and see if there seems to be a relationship between the two variables. (Go ahead, I'll give you 30 seconds.)

I'd say there is a pretty clear relationship there. The people who scored the highest on the cheese scale also scored the highest on the cheesecake scale. The groups of people didn't score exactly the same on both scales, of course, and the rank order isn't even the same, but, relatively speaking, the position of each person to each of the other people when it comes to cheese attitude is about the same as when it comes to cheesecake attitude. The Association's marketer has support for her hypothesis.

Computing a Correlation Coefficient

Just eyeballing two columns of numbers from a sample, though, is usually not enough to really know whether there is a relationship between two things. The marketing specialist in our example wants to use a single number to more precisely describe whatever relationship is seen.

The *correlation coefficient* takes into account all the information we used when we looked at our two columns of numbers in Table 2-1 and decided whether there was a relationship there. The correlation coefficient is produced through a formula that does the following things:

1. Looks at each score in a column
2. Sees how distant that score is from the mean of that column
3. Identifies the distance from the mean of its matching score in the other column
4. Multiplies the paired distances together
5. Averages the results of those multiplications

If this were a statistics textbook, I'd have to present a somewhat complicated formula for calculating the correlation coefficient. To call it *somewhat* complicated is generous. Frankly, it is terrifically frightening. For your own sanity, I'm not even going to show it to you. Trust me. Instead, I'll show you this pleasant, friendly looking formula (which works just as well):

$$\frac{\sum(Z_x Z_y)}{N-1}$$

Z refers to a *Z-score*, which is the distance of a score from the mean. These distances are then divided by the standard deviation for that distribution. So, Zx means all the Z-scores from the first column, and Zy means all the Z-scores from the second column. $ZxZy$ means multiply them together. The Σ symbol means *add up*. So, the equation says to multiply together all the pairs of Z-scores and add those cross-products together. Then, divide by the number (N) of pairs of scores minus 1.

The *mean* is the arithmetic average of a group of scores. It is produced by adding up all the numbers and dividing by the number of scores. A *standard deviation* for a group of numbers is the average distance of each score from the mean.

Before I produce the *Z-scores* used in our correlation formula, I need to know the means and standard deviations for each column of data. Equations for calculating these key values are provided in "Describe the World Using Just Two Numbers" [Hack #2]. Here are the means and standard deviations for our two variables:

Attitude toward cheese
    Mean = 33; standard deviation = 15.65

Attitude toward cheesecake
    Mean = 27.6; standard deviation = 7.44

Table 2-2 shows some of the calculations for our cheese attitude data.

Table 2-2. Calculations for discovering relationship between cheese attitude and cheesecake attitude

Person	Attitude toward cheese	Attitude toward cheesecake	Z-scores for cheese	Z-scores for cheesecake	Cross products of Z-scores
Larry	50	36	1.09	1.13	1.23
Moe	45	35	.77	.99	.76
Curly Joe	30	22	–.19	–.75	.14
Shemp	30	25	–.19	–.35	.07
Groucho	10	20	–1.47	–1.02	1.50

The correlation is .93. This is very close to 1.0, which is the strongest a positive correlation can be, so the cheese-to-cheesecake correlation represents a very strong relationship.

Interpreting a Correlation Coefficient

Somewhat magically, the correlation formula process produces a number, ranging in value from −1.00 to +1.00, that measures the strength of relationship between two variables. Positive signs indicate the relationship is in the same direction. As one value increases, the other value increases. Negative signs indicate the relationship is in the opposite direction. As one value increases, the other value decreases. An important point to make is that the correlation coefficient provides a standardized measure of the strength of linear relationship between two variables [Hack #12].

The *direction* of a correlation (whether it is negative or positive) is the artificial result of the direction of the scale one chooses to use to measure the variables. In other words, strong correlations can be negative. Think of a measure of golf skill correlated with average golf score. The higher the skill, the lower the score, but you would still expect a strong relationship.

Statistical Significance and Correlations

Our marketing specialist is likely also interested in whether a sample correlation is large enough that it is likely to have been drawn from a population where the correlation is bigger than zero. In other words, is the correlation we found in our sample so large that it must have come from a population where there is at least some sort of relationship between these variables?

The marketer in our example trusts correlations between a large number of pairs more than she does correlations from a small sample (such as our five bus riders). If she were to report this relationship to her boss and it wasn't true about most people, she might find herself selling cheesecake out of her minivan for a living.

Table 2-3 shows how large a correlation in a sample must be before statisticians are sure that there is a relationship greater than zero in the population it represents.

Table 2-3. Correlations that likely did not occur by chance

Sample size	Smallest correlation considered statistically significant
5	.88
10	.63
15	.51
20	.44
25	.40
30	.38

Table 2-3. Correlations that likely did not occur by chance (continued)

Sample size	Smallest correlation considered statistically significant
60	.26
100	.20

With our sample of five people, any correlation at least as big as .88 would be treated as *statistically significant* (which means "so big it probably exists in whatever population you took your sample from").

Where Else It Works

You can produce a correlation coefficient as a measure of the strength of a relationship between any two variables as long as certain conditions are met:

- You must be able to measure the variables in a way where numbers have real meaning and represent some underlying continuous concept. Examples of *continuous* variables are attitude, feelings, knowledge, skill, and things you can count, such as pounds gained because of love of cheesecake. (If the thing you are measuring is not continuous, as in the case when you have different categories, such as gender or political party, you can still calculate a correlation, just not with the formula shown here.)

- The variables must actually vary. If everyone felt the same about cheese, you couldn't calculate a correlation with attitude towards cheesecake or chocolate or anything. The math requires some variability.

- The minimum correlation sizes required to have statistical significance (shown in Table 2-3) are precisely accurate only when the sample is randomly drawn from the population. Researchers, such as our cheesecake marketer, must decide whether their sample is representative in the way a random sample would be.

Dire Warning About Correlations

It's tempting to treat correlational evidence as evidence of cause and effect. Of course, there might be all sorts of reasons why two things are related that have nothing to do with one thing causing the other.

For example, in the presence of such a strong correlation between attitude toward cheese and attitude toward cheesecake, you might want to conclude that a person's affinity for cheese *causes* him to like cheesecake because there is cheese in it. There might be noncausal explanations, though. The same people who like cheese might tend to like cheesecake because they like all foods that are kind of soft and smooshy.

Graph Relationships

Whenever a relationship between two variables is discovered and defined, we can use one variable to guess another. Drawing a regression line allows you to picture the relationship and make predictions.

So, you've just been named assistant regional manager of ice cream sales for 10,000 square feet of prime beachfront retail space along the shores of Sunflower Lake in northeast Kansas. Congratulations! You have a lot of responsibility and many strategic decisions to make about how to maximize profit. One dilemma that you will confront is whether to even open. Being open costs money and uses resources, and if you will sell few ice cream cones that day, it probably won't be worth it to even unlock the service window of your brightly painted plywood shack.

If only there were some way to magically know how good business will be on any given day. As an amateur statistician, you assume there must be a scientific way to guess how many cones will sell without having to actually open for business and test the market for the day. You're in luck. There *is* a way to make estimates of the value or score on some variable (such as ice cream sales) by using other information.

The key is that the other information must come from a variable that is related to the variable of interest. By drawing a line that shows the relationship among your variables for the days you know, you can look at the line as it extends into the future (or the past) for the days you do not know and guess what will happen. Such a graphic tool is called a *regression line*.

Drawing a Picture of the Future

Observant folks often discover correlations between variables [Hack #11]. The usefulness of knowing that a relationship exists goes beyond descriptive statistics, however.

Imagine that you have data on the activities around Sunflower Lake. Among other things, you have collected information about the amount of ice cream sales under the former assistant regional manager of ice cream sales (in number of ice cream cones sold) and the high temperature for each day (in degrees Fahrenheit). The correlation coefficient that represents the relationship between heat and craving for ice cream should be positive and fairly large. That is, as the heat increases, sales probably increase.

Intuitively, it makes sense that with some experience, you could look at the thermometer and get a sense of how busy the ice cream stand is going to be that day. Once you know that there is a positive or negative relationship between two variables, it makes sense that knowing the score on one will give you a general idea of what the score is on the other.

Once you find a relationship between two variables like this, it is reasonable to assume that the relationship between your two variables is *linear*. In other words, if you produce a graph with all the possible values of one variable as the X-axis (the horizontal line along the bottom) and all the possible values of the other variable as the Y-axis (the vertical line along the side) and then plot each pair of scores, the resulting dots form an essentially straight line.

Connecting the Dots

Figure 2-1 shows a way to graph the relationship between the temperature and ice cream sales at the beach.

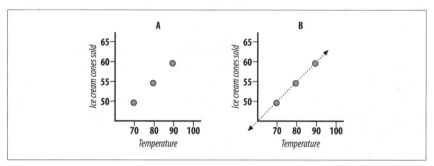

Figure 2-1. A linear relationship between sales and temperature

Graph A places dots to represent both values on the two variables, based on historic information you have collected. For instance, the lowest dot means that at 70 degrees, 50 ice cream cones were sold. At 90 degrees, 60 cones were sold. There is a clear pattern here, and the relationship looks like a straight line. For every 10-degree jump in temperature, sales go up 5 cones. For every 1-degree change in temperature, there is a $1/2$-cone increase in sales. Graph B draws a line based on this rule. The line goes through every dot.

In Figure 2-1, analyze Graph B to get a sense of the power of a regression equation. The line includes territory that is not sampled by the data. For instance, we do not have data for 100-degree days. With the regression equation, though, we can estimate what sales might be. If we place a dot on the line at the 100-degree mark, it appears to match up with the 65-cones mark. Using this regression equation, we could estimate that on 100-degree days, 65 cones would be sold. We could do the same for cooler days. Our graph suggests that on a 60-degree day, 45 cones would be sold.

Playing "What If?"

The relationship between heat and cone sales can be expressed mathematically. Our data for graphs A and B in Figure 2-1 look like this:

High temperature	Ice cream cones sold
70	50
80	55
90	60

So, let's see how we could build an equation that describes the relationship using numbers. Regression lines are statistical tools, after all. Notice that if we start with 70 degrees, we get 50 cones. If we enter 70 into our formula, we want 50 to be the output. We also want 80 to get us 55, and 90 to get us 60.

I played around with different possibilities using these values in an attempt to figure out what must be done to the input number to get the correct output number. I noticed that the "ice cream cones sold" value was always smaller than the temperature variable, so I wanted an equation that would shrink the temperature. Linear equations require a constant (some value to use in every equation) in order to produce a straight line, so I needed to have a constant in my equation as well. Rather than use trial and error, you could also enter this data into a statistics program, such as SPSS, or a spreadsheet, such as Excel, to produce the correct components. I found that this formula works well:

Cones Sold = 15 + (Temperature × .50)

> Algebraically, if you begin with a constant and then add some standard amount that is altered only through basic mathematical functions, such as multiplication, you will define a straight line that can be graphed.

"What if?" is a fun game to play with regression lines. Enter a value in one end and a guess comes out the other end; you can get an answer even for unrealistic scenarios. Throw some crazy value onto the line, such as 200 degrees, and you can still get an estimate for cone sales: 115!

The regression equation for this relationship would describe a line that could be drawn to show this relationship visually. With real data, the relationship is seldom as clear as it is in our example. (The correlation for our small fictional data set is a perfect 1.0.)

> In statistics, regression formulas make use of the correlation coefficient, the means, and the standard deviations of both groups of variable scores, regardless of the strength of the relationship in the data set. "Use One Variable to Predict Another" [Hack #13] presents statistical methods for producing a regression equation.

Why It Works

The accuracy of these sorts of regression estimates depends on a couple important factors. First, the relationship between variables must be fairly large. Small relationships produce dots all over the place in patterns that aren't straight at all, and a regression line drawn through such a mess misses a lot of dots and is not accurate. Unfortunately, in the social sciences, we don't find very many really strong relationships, so regression predictions tend to produce a certain number of errors. In statistics, errors come with the territory.

Second, the relationship must be at least *sort of* linear. As in our ice cream cone example, if the nature of the relationship changes somewhere along the regression line, the regression line will miss some of the data. Fortunately, most relationships in the natural world are linear or at least close to it.

Where It Doesn't Work

The actual relationship might not be exactly linear, but if it is essentially so, then regression analysis works pretty well. For example, with our ice cream example, maybe there is a certain increase in sales for every degree jump in the temperature. If that increase is the same regardless of where we are on the scale, we'll see a linear relationship. It is possible, though, that sales jump once a certain temperature is reached. Perhaps once it is over 90 degrees at the beach, people really flock to get relief.

Graphs C and D in Figure 2-2 show what happens if the true relationship isn't exactly linear.

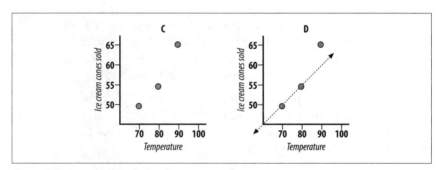

Figure 2-2. A nonlinear relationship

Following the requirements of linear regression, the regression equation always produces a straight line and, in this case, two of the dots fall right on it, but one does not. This line does a decent job of explaining the data by picturing the relationship, but because the relationship is not linear, the regression equation makes some errors.

Use One Variable to Predict Another

HACK #13

Simple linear regression is a powerful tool for measuring something you cannot see or for predicting the outcome of events that have not happened yet. With some help from our special friend statistics, you can make a precise guess of how someone will score on one variable by looking at performance on another.

Many professionals, both in and outside of the social sciences, often need to predict how a person will perform on some task or score on some variable, but they cannot measure the critical variable directly. This is a common need when making admission decisions into college, for example. Admissions officers want to predict college performance (perhaps grade point average or years until completion). However, because the prospective student has not actually gone to college yet, admissions officers must use whatever information they can get now to guess what the future holds.

Schools often use scores on standardized college admissions tests as an indicator of future performance. Let's imagine that a small college decides to use scores on the American College Test (ACT) as a predictor of college grade point average (GPA) at the end of students' first years. The admissions office goes back through a few years of records and gathers the ACT scores and freshman GPAs for a couple hundred students. They discover, to their delight, that there is a moderate relationship between these two variables: a correlation coefficient of .55.

Correlation coefficients are a measure of the strength of linear relationships between two variables [Hack #11], and .55 indicates a fairly large relationship. This is good news because the existence of a relationship between the two makes ACT scores a good candidate as a predictor to guess GPA.

Simple linear regression is the procedure that produces all the values we need to cook up the magic formula that will predict the future. This procedure produces a regression line that we can graph to determine what the future holds [Hack #12], but once we have the formula, we don't actually need to do any graphing to make our guesses.

Cooking Up the Equation

First, examine the recipe for creating the formula (see the "Regression Formula Recipe" sidebar), and then we'll see how to use it with real data. You can clip this recipe out and keep it in the kitchen drawer.

Regression Formula Recipe

Ingredients

2 samples of data from correlated variables:

- 1 criterion variable (the one you want to predict)
- 1 predictor variable (the one you will predict with)

1 correlation coefficient of the relationship between the 2 variables

2 sample means

2 sample standard deviations

Container

An empty equation shaped like this:

$$\text{Criterion} = \text{Constant} + (\text{Predictor} \times \text{Weight})$$

Directions

Calculate the weight by which you will multiply your predictor variable:

$$\text{Weight} = \text{correlation coefficient} \frac{\text{Criterion Standard Deviation}}{\text{Predictor Standard Deviation}}$$

Calculate the constant:

$$\text{Constant} = \text{Criterion Mean} - (\text{Weight} \times \text{Predictor Mean})$$

Fill the regression equation with the weight and constant you just prepared.

Serves

Anyone interested in guessing *what would happen if....*

The regression recipe calls for two other ingredients, means and standard deviations for both variables. Here are those statistics for our example:

Variable	Mean	Standard deviation
ACT scores	20.10	2.38
GPA	2.98	.68

You can review means and standard deviations in "Describe the World Using Just Two Numbers" **[Hack #2]**.

The admissions office built a regression equation from this information. Consequently, as each applicant's letter came into the admissions office, an officer could enter the student's ACT score into the regression formula and predict his GPA. Let's figure out the parts of the regression equation in this example:

$$\text{Weight} = \text{correlation coefficient} \frac{\text{Criterion Standard Deviation}}{\text{Predictor Standard Deviation}}$$

$$\text{Weight} = .55\frac{.68}{2.38} \qquad \text{Weight} = .55(.29) \qquad \text{Weight} = .16$$

$$\text{Constant} = \text{Criterion Mean} - (\text{Weight} \times \text{Predictor Mean})$$

$$2.98 - (0.16 \times 20.10) = 2.98 - 3.22 = -.24$$

By placing all this information into the regression equation format, we get this formula for predicting freshman GPA using ACT scores:

$$\text{Criterion} = \text{Constant} + (\text{Predictor} \times \text{Weight})$$

$$\text{Predicted GPA} = -.24 + (\text{ACT score} \times .16)$$

> Notice that the constant in this case is a negative number. That's OK.

Predicting Scores

In our college admissions example, imagine two letters arrive. One applicant, Melissa, has an ACT score of 26. The other applicant—let's call him Bruce—has an ACT score of 14.

Using the regression equation we have built, there would be two different predictions for these folks' eventual grade point averages:

For Melissa

- Predicted GPA = $-.24 + (26 \times .16)$
- Predicted GPA = $-.24 + 4.16$
- Predicted GPA = 3.90

For Bruce

- Predicted GPA = $-.24 + (14 \times .16)$
- Predicted GPA = $-.24 + 2.24$
- Predicted GPA = 2.00

I hope, for Bruce's sake, there is more than one spot available.

The two variables in this example, ACT scores and GPA, are on different scales, with ACT scores typically running between 1 and 36 and GPA ranging from 0 to 4.0. Part of the magic of correlational analyses is that the variables can be on all sorts of different scales and it doesn't matter. The predicted outcome somehow knows to be on the scale of the criterion variable. Kind of spooky, huh?

Why It Works

When two variables correlate with each other, there is overlap in the information they provide. It is as if they share information. Statisticians sometimes use correlational information to talk about variables *sharing variance*.

If some of the variance in one variable is accounted for by the variance in another variable, it makes sense that smart mathematicians can use one correlated variable to estimate the amount of variance from the mean (or distance from the mean) on another variable. They would have to use numbers that represent the variables' means and variability, and a number that represents the amount of overlap in information. Our regression equation uses all that information by including means, standard deviations, and the correlation coefficient.

Where Else It Works

Regression is helpful in answering research questions beyond making predictions. Sometimes, scientists just want to understand a variable and how it operates or how it is distributed in a population. They can do this by looking at how that variable is related to another variable that they know more about.

Statisticians call simple linear regression *simple* not because it is easy, but because it uses only one predictor variable. It is simple as compared to *complex*. Real-life predictions like those in our example usually use many predictors, not just one. The method of predicting a criterion variable using more than one predictor is called multiple regression [Hack #14].

Where It Doesn't Work

There will be error in predictions under three circumstances. First, if the correlation is less than perfect between two variables, the prediction will not be perfectly accurate. Since there are almost never really large relationships between predictors and criteria, let alone perfect 1.0 correlations, real-world applications of regression make lots of mistakes. In the presence of any correlation at all, though, the prediction is more accurate than blind guessing. You can determine the size of your errors with the standard error of estimate [Hack #18].

Second, linear regression assumes that the relationship is linear. This is discussed in "Graph Relationships" [Hack #12] in greater detail, but if the strength of the relationship varies at different points along the range of scores, the regression prediction will make large errors in some cases.

Finally, if the data collected to first establish the values used in the regression equation are not representative of future data, results will be in error. For example, in our college admissions example, if an applicant presents with an ACT score of 36, the predicted GPA is 5.52. This is an impossible value that does not even fit on the GPA scale, which maxes out at 4.0. Because the past data that was used to establish the prediction formula included few or no ACT scores of 36, the equation was not equipped to deal with such a high score.

Use More Than One Variable to Predict Another

The super powers of predicting the future and seeing the invisible are available to any statistics hackers who feel they are worthy. Statisticians often answer questions and use correlational information to solve problems by using one variable to predict another. For more accurate predictions, though, several predictor variables can be combined in a single regression equation by using the methods of multiple regression.

"Graph Relationships" [Hack #12] discusses the useful prophetic qualities of a regression line. Those procedures allow administrators and statistical researchers to predict performance on assessments never taken, understand variables, and build theories about relationships among those variables. They accomplish these tricks using just a single predictor variable.

"Use One Variable to Predict Another" [Hack #13] presents the problem colleges have when deciding which applicants to admit. They want to admit students who will succeed, so they try to predict future performance. The solution in that hack uses one variable (a standardized test score) to estimate performance on a future variable (college grades).

Often, real-life researchers want to make use of the information found in a bunch of variables, not just one variable, to make predictions or estimate scores. When they want greater accuracy, scientists attempt to find several variables that all appear to be related to the criterion variable of interest (the variable you are trying to predict). They use all this information to produce a *multiple regression equation*.

Choosing Predictor Variables

You probably should read or reread "Use One Variable to Predict Another" [Hack #13] before going further with this hack, just to review the problem at hand and how regression solves it. Here is the equation we built in that hack for using a single predictor, ACT scores, to estimate future college admission:

Predicted GPA = –.24 + (ACT Score × .16)

This single predictor produced a regression equation with output that correlated .55 with the criterion. Pretty good, and pretty accurate, but it could be better.

Imagine our administrator decides she's unhappy with the level of precision she could get using the regression line or equation she had built, and wants to do a better job. She could get a more accurate result if she could find more variables that correlate with college grades. Let's imagine that our amateur statistician found two other predictor variables that correlated with college performance:

- An attitude measure
- The quality of a written essay

Perhaps performance on a college attitude survey is collected by the college (scores range between 20 and 100), and is found to have some correlation with future GPA. Additionally, a score of 1 to 5 on a personal essay could correlate with college GPA and might be included in the multiple regression equation.

Building a Multiple Regression Equation

Let's look first at the abstract format of the regression equation in general. Then, we'll apply the tool to the task at hand. Here is the basic regression equation using just one predictor variable:

Criterion = Constant + (Predictor × Weight)

If you want to use more information, you can extend this equation to include more predictors. Here's an equation with three predictors, but you could expand the equation form to include any number of predictors:

Criterion = Constant +
     (Predictor 1 × Weight 1) +
     (Predictor 2 × Weight 2) +
     (Predictor 3 × Weight 3)

Each predictor has its own associated weight, which is determined through statistical formulas that are based on the correlation between the predictor and the criterion variable. The equations for this process are somewhat complex, so I won't show them here. (You're welcome.) In real-life regression equation building, computers are almost always used to produce multiple regression equations.

I used the statistical software SPSS for many of the computations in this book, using data, often fictional, that I entered into SPSS data files. Microsoft's Excel is another handy tool for performing simple statistical analyses.

Using realistic data that we might find with three predictors that correlate with the criterion, as well as correlate with each other somewhat, we might produce a regression equation with values like this:

Predicted GPA = 3.01 +
$$(\text{ACT Score} \times .02) +$$
$$(\text{Attitude Score} \times .007) +$$
$$(\text{Essay Score} \times .025)$$

With the imaginary data I used on my computer to produce these weights, the overall equation predicted college GPA very well, finding a correlation of .80 between observed GPA values and predicted GPA values. This is much better than the .55 correlation of our single predictor.

When we add two other predictors to the *model* (a description of a group of variables and how they are related), specifically the attitude measure and the essay score, the weight for the ACT score changes. This is because of the use of *partial correlations* instead of one-to-one correlations for each predictor. In addition, the constant changes. This is discussed later, in the "Why It Works" section of this hack.

Making Predictions and Understanding Relationships

To estimate what a prospective student's college performance will be, our administrator takes the scores for that student on each of the predictors and enters them into the equation. She multiplies each predictor score by its weight and adds the constant. The resulting value is the best guess for future performance. It might not be exactly right, of course (and, in fact, is most likely not exactly right), but it is a better guess than having no information at all.

If you have no information at all and have to guess how a student will do in college, you should guess that she will earn the mean GPA, whatever that is for your institution.

What if you want to do more than just predict the future, and want to really understand the relationships between your predictors and the criterion? You might do this because you want to build a more efficient formula that doesn't require a bunch of information that isn't very useful. You also might do it just because you want to build theory and understand the world, you crazy scientist, you! The problem is that it is hard to know the independent contribution of each predictor by just looking at the weights.

The weights for each variable in a multiple regression equation are scaled to the actual range of scores on each variable. This makes it hard to compare each predictor to figure out which provides the most information in predicting the criterion. Comparing these raw weights can be misleading, as a variable might have a smaller weight just because it is on a bigger scale.

Compare the weight for ACT score with the weight for attitude score, for example. The weight of .02 for ACT is larger than the weight of .007 for attitude, but don't be fooled into thinking that ACT scores play a larger role in predicting GPA than attitude. Remember, GPA scores range from about 1.0 to 4.0, whereas attitude scores range from 20 to 100. A smaller weight for attitude actually results in a bigger jump on the criterion than does the larger weight for ACT scores.

Computer program results for multiple regression analyses often provide information in the format shown in Table 2-4.

Table 2-4. Multiple regression results

Criterion	Nonstandardized weights	Standardized weights
Constant	3.01	-----
ACT scores	.02	.321
Attitude scores	.007	.603
Essay scores	.025	.156

The third column in Table 2-4 is more useful than the "Nonstandardized weights" values in identifying the key predictors and comparing the unique contributions of each predictor to estimating the criterion.

Standardized weights are the weights you would get if you first convert all the raw data into z scores [Hack #26]: the distance of each raw score from the mean expressed in standard deviations.

The standardized weights have placed all predictors on the same scale. By doing this, the relative overlap of each predictor with the criterion can be fairly compared and understood. For example, with this data, it is probably appropriate to say that attitude explains twice as much about college GPA than does ACT performance, because the standardized weight for attitude is .603, about twice as much as the standardized weight for ACT scores (.321).

Why It Works

Multiple linear regression does a better job in predicting outcomes than simple linear regression because multiple regression uses an additional bit of information to compute the exact weights for each predictor. Multiple regression knows the correlation of each predictor with the other predictors and uses that to create more accurate weights.

This bit of complexity is necessary because if the predictors are related to each other, they share some information. They aren't really independent sources of prediction if they correlate with each other. To make the regression equation as accurate as possible, statistical procedures remove the shared information from each predictor in the equation. This produces independent predictors that come at the criterion from different angles, producing the best prediction possible.

> Imagine two predictor variables that correlate perfectly with each other—that is, correlation equals 1.00. Using both variables in a regression equation would be no more accurate than using just one (doesn't matter which one) by itself. By extension, any overlap between predictors (i.e., any correlation between predictors greater or less than 0.00) is redundant information.

Figure 2-3 illustrates the use of multiple sources of independent information to estimate a criterion score.

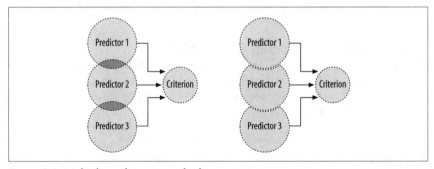

Figure 2-3. Multiple predictors in multiple regression

The correlation information used to determine the weight for each predictor in multiple regression is not the one-to-one correlation between a predictor and the criterion. Instead, it is the correlation between a predictor and the criterion when the overlap among all the predictors has been removed.

This process produces predictor variables that are somewhat different than the actual measure variables. By statistically removing (or *controlling* for) the shared information among predictors, the predictors are conceptually different than they were before. As Figure 2-3 shows, now they are independent predictors with a different "shape." The correlations between these altered variables and the criterion variable are used to produce the weights.

 Correlations between predictor variables and a criterion variable when all the redundant shared information has been statistically removed from the predictors are called *partial correlations*. Partial correlations are the one-on-one correlations you would get between each predictor and the criterion if the predictor variables do not correlate with each other.

Where Else It Works

Multiple regression is used every day by real people in the real world for one of two reasons. First, multiple regression allows for the construction of a prediction equation, so people can use scores on a group of variables that they have in front of them to estimate a score on another variable that they cannot have in front of them (because it is either in the future or cannot be measured easily for some reason). This is how the tool of multiple regression is used to solve problems in the world of *applied science*.

Multiple regression also allows for examination of the independent contribution that a group of variables make to some other variable. It allows us to see where there is information overlap among variables and build theories to understand or explain that overlap. This is how the tool of multiple regression is used to solve problems in the world of *basic science*.

H A C K #15 Identify Unexpected Outcomes

How do you know if your observations are correct or if you are just biased? How do you know when there is more or less of something than should have occurred by chance? You can find out for sure by using the flexible one-way chi-square test.

In science, the oldest type of observational research involved counting people, animals, and things:

- How many people are on this boat?
- What proportions of butterflies have little green spots on their wings?

As the field of inferential statistics matured, the questions became more specific:

- Were an equal number of boys and girls born in London in 1812?
- Are an equal number of crimes committed at different times of day?

The research question for these situations is "are they equal?" (or, at least, are they close enough that any fluctuations are probably due to chance). The implication of an unequal distribution is that *something is going on*. What, exactly, is going on cannot be answered by this sort of question. It is a start, though, and a good first question to ask.

Have you ever noticed that *something* seemed to be *going on*, but weren't sure if it was just your imagination? Do a greater number of hippies shop at the local community mercantile than would be expected by chance? If the answer is yes, and you are looking to meet hippies, you should start hanging out there.

In business, and for those who have to provide services, identifying where there is the most need is crucial. Observational data can be used to solve that problem. Even just in everyday life, we all have our beliefs (which might be biased) that are based on observations. I *have* noticed a lot of hippies at the community mercantile, but maybe I am just on the lookout for hippies when I am in that store. Are there really more hippies than normal there? More hippies than, say, nonhippies?

These sorts of questions can be answered using a statistical tool appropriate for seeing whether the number of "things" within each of a number of categories is more unequal than would normally be found by chance. This tool is named the *one-way chi-square*.

> This statistical analysis is called chi-square because the symbol used for the critical value generated is an X, which is the Greek letter *chi* (pronounced "kye"). The values needed in the calculations are all squared, thus we call this whole thing a *chi-square* or *chi-squared*.

Determining Whether Something Is Going On

Imagine you are responsible for scheduling the police officers in your town. The problem is that you don't know whether to schedule the same amount of officers for every shift or whether more crime might occur during particular shifts. If one shift is likely to be busier, you should probably assign more officers. Of course, another reason to assign more officers during that time is that their patrolling might cut crime down a bit.

Here is an example of some imaginary data describing crime events for three periods of time. Imagine the data was collected over a 30-day period, and you would like to use this data to plan for the coming year. The numbers indicate how many crimes were committed during each of three police shifts.

Midnight – 8 a.m.	8 a.m. – 4 p.m.	4 p.m. – Midnight	Total
120	90	90	300

It certainly looks like more crimes occur late at night. By observation alone, we might conclude that there is more crime late at night. Perhaps that is just in our sample, though, and there really isn't a difference in the population of all the data we could have collected.

Calculating the Chi-Square

We could compute a chi-square for this data. If the chi-square is really big, then the 120 crimes is unusually larger than the other two crime periods. How big "really big" needs to be is an important question that we will explore later in this hack.

Here's how to think about the analysis we are about to do. If there are 300 crimes committed in one 24-hour period, we would expect 33.3 percent of them, or 100, to occur in each of three equally long time intervals during the day. If there is more or less than 100 for each of those intervals, *something is going on*. Perhaps the time of day matters in the commission of crimes. Of course, there might be some chance fluctuation, but the larger the difference between the expected and the actual frequencies, the less likely that those differences are just chance.

Here is the chi-square formula:

$$\text{Chi-square} = \sum \frac{(\text{Observed Frequency} - \text{Expected Frequency})^2}{\text{Expected Frequency}}$$

Σ is a symbol that means to sum or add up the things that follow it.

Let's calculate a chi-square for this data. The observed frequency for each category is given. The expected frequency for each cell would be 300 divided by three categories, or 100:

$$\frac{(120-100)^2}{100} + \frac{(90-100)^2}{100} + \frac{(90-100)^2}{100}$$

$$\frac{(20)^2}{100} + \frac{(-10)^2}{100} + \frac{(-10)^2}{100} = \frac{400}{100} + \frac{100}{100} + \frac{100}{100} = 4 + 1 + 1 = 6$$

The chi-square for this data is 6. Okay. Now what? Is 6 big or small or what? Could a chi-square as big as 6 occur by chance?

Determining if the Chi-Square Is "Really Big"

As with all statistics—such as correlation coefficients [Hack #11], t tests [Hack #17], proportions, and so on—statisticians have mapped out the distribution of the chi-square. In other words, we know the likelihood that chi-squares of different sizes will occur by chance. The likelihood of finding chi-squares of particular magnitudes depends on the number of categories.

Table 2-5 shows a portion of a theoretically giant table that shows the chi-square values that one must beat in order to be 95 percent sure (level of significance = .05) that the value didn't get that big just because of chance fluctuations in the sample. We know these critical values occur by chance 5 percent or less of the time because chi-squares, like almost everything else in the orderly world of statistics, have a known distribution—i.e., a known set of likelihoods that certain values will occur. Like the normal curve, the chi-square distribution is well-defined [Hack #23].

Table 2-5. Critical chi-square values at the .05 level of significance

Two categories	Three categories	Four categories	Five categories
3.84	5.99	7.82	9.49

Our chi-square value is 6, which is higher than the critical value for three categories (5.99). This means something very specific, so I'll emphasize it. Though I am specifically referring to the crime rate problem at hand, I am using the same pattern of words that describe all statistical findings that are *significant at the .05 level*.

 If, in the population, there are no differences in the number of crimes committed at the three times of day, you would occasionally draw out random samples with differences that produce a chi-square of 6 or larger, but it would happen less than 5 percent of the time.

It seems reasonable to conclude, then, that in the population there *are* differences in frequency of crime based on time of day. Because these differences are "real," it is reasonable to schedule a year's worth of police patrols based on them.

Why It Works

Data for chi-square analyses are laid out in a way in which the observed number of things in each category can be compared with the expected number of things in each category. The "expected number of things in each category" is usually defined as an equal number. If nothing is going on (i.e., if the category makes no difference), we expect an equal number of things in each category.

Chi-squares work with categorical data. Essentially, the difference between what was expected and what was observed is computed for each category. The differences are compared to the expected frequency (as a way to standardized all the differences), and then those ratios are all added together. The size of the resulting number determines its likelihood of occurring by chance. The bigger the number, the less likely that chance alone explains things. There is a known *distribution* (list of probabilities associated with each possible chi-square value) that is used by a table (or computer) to assign a specific probability to each chi-square value.

If there are two or more categories and the researcher wants to know whether the actual distribution across these categories is what would be expected by chance alone, then the chi-square is an appropriate test. The actual value that is tested is the difference between what the researcher expects to find and what actually occurs.

The chi-square test is used in the framework of having certain *expectations* and seeing whether they are met by the *observed* data. This is a simple form of model testing. The researcher has a belief system, in the form of some model or hypothesis of how the world should behave. She then observes the world (collects data) and compares her observations to her model. If the data fits the model, this is support for her hypotheses. The chi-square test, consequently, is considered a *goodness-of-fit statistic*. It answers the question of how well the data fits a model.

> Some statistics textbooks refer to the *one-way chi-square* as the *single sample chi-square*, so don't get confused. But what are you doing reading some other statistics book anyway?

Statisticians know the size of normal fluctuations in observed frequencies compared to expected frequencies. With this knowledge, they can compute the likelihood that any observed deviation from the expected occurs by chance or because something else is going on.

Where Else It Works

Though a simple and historically ancient (about 80 years old, which is old by statistics standards!) statistical method, the chi-square is very useful for a variety of statistical questions at both low levels of measurement and, surprisingly, very advanced statistical methods. Because it is a fairly straightforward way to model test (or quantify "goodness of fit"), the chi-square is used as part of complex correlational analyses and measurement diagnostics.

Chi-square analyses are used to see whether complicated theoretical models of the world—comprehensive maps of relationships among variables—actually match real-world data. If the real world deviates too much from the expectations implied by one of these models, it is concluded that the model is weak. A significant chi-square is the criterion used for "too much" deviation.

For example, if test developers are concerned about item bias (that one item might work differently for one identifiable group over another—such as races, genders, and so on), they will check whether the patterns of answer options meet certain expectations regardless of which group generated the data. The chi-square analysis compares the expectations to actual test performance.

See Also

- "Identify Unexpected Relationships" [Hack #16]

Identify Unexpected Relationships

HACK #16

If you want to verify whether a relationship you have observed between two variables is real, you have a variety of statistical tools available. A problem arises, though, when you have measured these variables without much precision, using categorical measurement. The solution is a two-way chi-square test, which, among other things, can be used to make unsubstantiated assumptions about the characteristics of people you have just met.

"Identify Unexpected Outcomes" [Hack #15] used the *one-way chi-square test* to make police scheduling decisions based on whether equal numbers of crimes were committed at different times of day. That tool works well to solve any analytical problem when:

- The data is at the categorical level of measurement (e.g., gender, political party, ethnicity).

- You want to determine whether there is a greater frequency of scores in certain categories than would be expected by chance.

You face another common analytic problem when you're curious to know whether two categorical variables are *related* to each other. Relationships between categorical variables can be examined with the handy *two-way chi-square* test.

> If two variables are measured at the interval level (many scores are possible along a continuum), the correlation coefficient [Hack #11] is the best tool to use, but it doesn't work well with categorical measurement.

We make assumptions all the time about relationships between these sorts of variables. Many of our common stereotypes about categories of people have implicit hypotheses about these relationships. Here are a few assumptions you might have that imply a relationship between categorical variables:

- Professors are absent-minded.
- Computer programmers play *Dungeons and Dragons*.
- Adults who collect comics write *Statistics Hacks* books.
- Professors are absent-minded.

If you meet a computer programmer at a party and you hold this stereotype belief about this type of person, you might assume that she is familiar with 20-sided dice. If you are wrong, though, that might lead to much awkward conversation. It would be nice to know if there really were such relationships between these categorical variables of interest. Calculating a two-way chi-square solves this problem and can verify or cast doubt on these assumptions about people.

Answering Relationship Questions

While the one-way chi-square analyzes a single categorical variable, two-way chi-squares analyze the relationship between *two* categorical variables. The process is the same: compare the expected frequencies with actual frequencies for each category or combination of categories. If the differences add up to a big number, then *something is going on*.

Here is a categorical relationship question that we might like to have answered. It is similar to other issues of stereotype that could be explored:

Are females more likely to be Democrats or Republicans?

You probably already have some assumption about this, but how would you go about checking the accuracy of such an assumption?

Review of the One-Way Chi-Square

The chi-square test is used in the framework of having certain *expectations* and seeing whether they are met by the *observed* data. Statisticians know the size of normal fluctuations in observed frequencies compared to expected frequencies. With this knowledge, they can place a likelihood that any observed deviation from the expected occurred by chance or whether something else is going on. The raw data for these analyses is usually the number of people (the frequency) in each category of some variable.

Here is the general chi-square formula:

$$\text{Chi-square} = \sum \frac{(\text{Observed Frequency} - \text{Expected Frequency})^2}{\text{Expected Frequency}}$$

Σ means to add up the things that follow it. The bigger the chi-square, the less likely it is that the outcomes occurred randomly.

Conduct preliminary analyses. Look at Table 2-6 for an example of categorical frequency data for, to start, a single categorical variable. This data is fictional, but consistent with published studies, which typically find that Republicans are more likely to be male and that females tend to more commonly identify as Democrats.

Table 2-6. Hypothetical sample of Republicans

Males	Females
45	30

In this random sample of 75 Republicans, 45 are males and 30 are females. That's 60 percent male and 40 percent female. Can we conclude that Republicans in general are more likely to be male than female? If not, we would expect there to be 50 percent males and 50 percent females in our sample.

A *one-way* chi-square could see whether more Republicans are male than female, but that's not the hack we are exploring here.

This isn't our research question, though.

Compute the two-way chi-square. Our initial question included only Republicans, so while political party might have seemed like a variable in our first analysis, it was really just a description of the population; it did not *vary* at

all. We can add *party* to our analysis, though, by adding another category—
Democrat, for example—and recruiting 75 more participants, and suddenly
we have data with two variables. Imagine frequency data as shown in
Table 2-7.

Table 2-7. Hypothetical sample of voters

Party	Males	Females	Totals
Republican	45	30	75
Democrat	34	41	75
Totals	79	71	150

Here we have two categorical variables: party affiliation and sex. We could
go ahead and use a one-way analysis to look at either of the two rows by
themselves. However, a more typical question would be, "Is there a relationship between party and sex?"

Q: "Is there a relationship between party and sex?"

A: Reminds me of my freshman year.

(Ha! I got a million of 'em. I'll be here all week. Good night,
everybody!)

To calculate a standardized measure of the difference between the expected
frequencies and the observed frequencies, we use the same formula as with
the one-way chi-square. As "Identify Unexpected Outcomes" **[Hack #15]** demonstrates, we start by totaling up the differences between expected and
observed frequencies in each *cell* (each square of a table).

We do the same with the two-way chi-square. The expected frequency in
each cell is equal to the number of people in that cell's row multiplied by the
number of people in that cell's column and then divided by the total sample
size. Using the data in Table 2-7, the calculations for expected frequencies
are shown in Table 2-8.

Table 2-8. Expected frequencies for two-way chi-square analysis

Party	Males	Females
Republican	$(75 \times 79) / 150 = 39.5$	$(75 \times 71) / 150 = 35.5$
Democrat	$(75 \times 79) / 150 = 39.5$	$(75 \times 71) / 150 = 35.5$

Thus, the two-way chi-square calculations look like this:

$$\text{Chi-square} = \frac{(45 - 39.5)^2}{39.5} + \frac{(34 - 39.5)^2}{39.5} + \frac{(30 - 35.5)^2}{35.5} + \frac{(41 - 35.5)^2}{35.5}$$

$$\text{Chi-square} = \frac{(5.5)^2}{39.5} + \frac{(-5.5)^2}{39.5} + \frac{(-5.5)^2}{35.5} + \frac{(5.5)^2}{35.5}$$

$$\text{Chi-square} = \frac{30.25}{39.5} + \frac{(30.25)}{39.5} + \frac{(30.25)}{35.5} + \frac{(30.25)}{35.5}$$

$$\text{Chi-square} = .77 + .77 + .85 + .85 = 3.24$$

Determine if the chi-square is big enough. Statisticians know that the critical chi-square value for 2×2 tables (like the chi-square we just computed) is 3.84. Chi-square values greater than 3.84 are found by chance about 5 percent of the time or less [Hack #15].

Because our chi-square value was 3.24 and that is less than the key 5 percent value of 3.84, we know that such a fluctuation can occur by chance somewhat greater than 5 percent of the time. We cannot claim statistical significance here, and so we must conclude that though our sample seemed to show a relationship between the two categorical variables of party affiliation and sex, it might have occurred because of chance sampling error. In the population from which the sample was drawn, there might not be any relationship.

Why It Works

A two-way chi-square answers this relationship question by looking at differences. This might seem counterintuitive, because most statistics look for differences in order to show, well, a difference, not to show similarities. But here's the thinking:

- If there is no relationship between party and sex, then each sex should be equally split between Republicans and Democrats.
- Also, if there is no relationship, then each party should be equally split between males and females.
- This equal distribution in both directions is what is expected by chance. Large deviations from those expectations suggest that *something is going on.*

The problem solved with this hack was one of knowing whether a stereotype belief we held was correct. Of course, outside of the real world, in the scientific world, researchers use this tool to explore a wide variety of complex questions.

Two-way chi-squares, sometimes called *contingency table analyses*, are useful anytime you have two categorical variables and want to see whether there is some dependency of one variable on the other. Our example used variables with just two categories, but similar analyses can be done on variables with many categories. The technical requirements are a bit more complex, but the procedure is the same.

See Also

- "Identify Unexpected Outcomes" [Hack #15]

H A C K   Compare Two Groups
#17
Which is better? Which has more? Do people really differ? Quantitative questions like these dominate the polite conversations of our times. If you want some real evidence for your beliefs about the best, most, and least, you can use a statistical tool called the "t test" to support your point.

My Uncle Frank is full of opinions. Green M&Ms taste better than blue. Women never get speeding tickets. The *Brady Bunch* kids could sing better than the *Partridge Family*. Plaid is back. He can argue all day spouting half-baked idea after half-baked idea. While I disagree with him on all four points (especially the position that plaid is *back*—after all, it never left!), I have only my opinions to fight with.

If only there were some scientific way to prove whether Uncle Frank is right or wrong! You no doubt recognize the rhetorical nature of my plea. After all, there are only about a gazillion statistical tools that exist to test hypotheses like these. One of the simplest tools is designed to test the simplest of claims. If the problem is deciding whether one group differs from another, the procedure known as an independent *t* test is the best solution.

Proving Uncle Frank Wrong (or Right)

To apply a *t* test to investigate one of Uncle Frank's theories, we have to compute a *t* value. Let's imagine that I decided to actually challenge Uncle Frank and collect some data to see whether he is right or wrong.

Uncle Frank believes that males get speeding tickets more frequently than females. To test this hypothesis, imagine that I select two groups of 15 drivers randomly [Hack #19] from his neighborhood. One group is female, and the other is male. I ask them some questions. Pretend that over the course of the last five years, the male group averaged 1.71 speeding tickets with a variance of .71. The female group averaged 1.35 speeding tickets with a variance of .25.

> *Variance* is the total amount of variability in a given group of numbers. It is calculated by finding the distance of each score in the group from the mean score. Square those distances and average them to get the variance.

Here is the equation for producing a *t* value:

$$t = \frac{\text{Mean of Group 1} - \text{Mean of Group 2}}{\sqrt{\dfrac{\text{Variance for Group 1}}{\text{Sample Size of Group 1}} + \dfrac{\text{Variance for Group 2}}{\text{Sample Size of Group 2}}}}$$

The larger the *t* value, the less likely that any differences found between your sample groups occurred by chance. Typically, *t* values larger than about 2 are big enough to reach the conclusion that the differences exist in the whole population, not just in your samples.

> The *t* formula shown here works best when both groups have the same number of people in them. A similar formula that averages the variance information is used when there are unequal sample sizes.

Is there support for Uncle Frank's belief? To determine that, our calculations require the data in Table 2-9.

Table 2-9. Data for speeding ticket t test

	Group 1 (males)	Group 2 (females)
Mean	1.71	1.35
Variance	.71	.25
Sample Size	15	15

If we place those key values into our *t* formula, it looks like this:

$$t = \frac{1.71 - 1.35}{\sqrt{\dfrac{.71}{15} + \dfrac{.25}{15}}}$$

The calculations work out this way:

$$t = \frac{.36}{\sqrt{.047 + .017}} = \frac{.36}{\sqrt{.064}} = \frac{.36}{.253} = 1.42$$

In this case, a mean difference of .36 produces a *t* value of 1.42.

Interpreting the t Value

Could our *t* value of 1.42 have occurred by chance? In other words, if the actual difference in the population is zero, could two samples drawn from that single population produce means that differ by that much?

Earlier, I mentioned that values of 2 or greater are typically required to reach this conclusion. Under this standard, we would conclude that there is no evidence that males really do get more tickets than females. They did in our sample, of course, but might not if we measured everybody (the whole population). There is no evidence that Uncle Frank is right. This is different in an important way from concluding that he is wrong, but it still means he should lose this particular argument.

Statistics is all about precision, though, so let's explore our 1.42 a little further. How big, exactly, would it need to be for us to conclude that Uncle Frank is actually right?

The answer, determined through custom, is that if the t is bigger than would occur by chance 5 percent of the time or less, then the t is big enough. Fortunately, the chances of finding ts of various sizes when drawn randomly from a population has been determined by hard-working mathematicians using assumptions of the Central Limit Theorem [Hack #2]. The exact t value required for statistical significance depends on the total sample size in both groups combined. Table 2-10 provides t values that you must meet or beat to declare statistical significance at the .05 level.

Table 2-10. t values occurring by chance less than 5 percent of the time

Sample size in both groups combined	Critical t value
4	4.30
20	2.10
30	2.05
60	2.00
100	1.99
∞(infinity)	1.96

For sample sizes other than those shown in Table 2-10, you can figure out the rough t value you need to meet or beat by estimating the value between the values shown. Also, the chart assumes that you want to identify differences between groups in either direction. It assumes you want to know whether either group mean is larger than the other. This is what statisticians call a *two-tailed test*, and it is usually the comparison of interest.

Using Table 2-10, we see that a t value of 1.42 is less than the critical value for a total of 30 subjects. We need to see a t value greater than 2.05 to be confident that the sample differences we observed did not occur just by chance.

Why It Works

Social scientists use this comparison method all the time. Experimental and quasi-experimental designs often have two groups of people who are believed to be different in some way or another. You might be interested in the differences between Republicans and Democrats or girls and boys, or you might want to see if a group taking a new drug has fewer colds than a group not taking a drug at all.

Such designs produce two sets of scores, and those sets of values often differ, at least in the samples used. Researchers (and I, too, when it comes to proving Uncle Frank wrong) are more interested in whether there would be differences in the *populations* represented by the two samples.

> The logic of inferential statistics is that a sample of scores represents a larger population of scores. If the samples differ on some variable, that difference might be reflected in the populations from which they were drawn. Or that difference might be due to errors resulting from the sampling.

A *t* test answers the question of whether any differences found between two samples are real (i.e., they probably exist in the populations from which the samples were drawn) or due to sampling error (i.e., they probably exist only in the samples). If the difference between the samples is too large to have occurred by chance, researchers conclude that there is a real difference between the populations.

The *t* test formula uses information about the shape of the sample distributions of scores. The needed information is the mean score on the research variable in each group, each group's variance, and the sample size of each group. The sample mean provides a good guess as to the population mean, the variances give an indication as to how much the sample mean might have varied from the population mean, and the sample size suggests the precision of the estimate. The difference between the two means is standardized and is expressed as a *t* value.

> The way statisticians talk about real differences is "the two samples were likely drawn from different populations." The way you and I and researchers might talk about real differences is "Republicans and Democrats differ" or "the drug reduces the chance of getting a cold."

Where Else It Works

Numbers don't know where they come from. You can use *t* tests to look at differences in any two sets of numbers, whether those numbers describe people or things. In fact, the *t* test was first developed to determine the quality of an entire elevator full of grain used in beer production.

Instead of examining all the grain, a beer statistician (how's that for a dream job?) wanted a method that requires looking at a small sample only, randomly drawn from the larger population of grain. The rest is history, and so we can say today that much of the work done by statistical researchers is literally driven by beer.

HACK #18 Find Out Just How Wrong You Really Are

Anytime you have used statistics to summarize observations, you've probably been wrong. If you need to know how close you have come to the truth, use standard errors.

Statisticians are perhaps the only professionals who not only proudly admit that their answers are probably wrong, but will go to great lengths to tell you exactly how wrong they are. When you conduct a survey, record observations, or conduct some sort of experiment, your results describe only your *sample*—the customers, patients, students, goldfish, or pieces of Kryptonite that you have in front of you. Inferential statistics uses values computed for a sample to estimate what that value would be for the *population* it is meant to represent. For example, the mean of a sample is a pretty good guess for the mean of the population. The problem is knowing whether to trust your results.

Calibrating Error and Calculating Precision

It is unlikely that the mean of a sample is exactly the same as the mean of the population, but it is likely to be close. If you want to know how far wrong you are, you can calibrate your precision using *standard errors*. The standard error of the mean gives us an estimate of the distance between our sample mean estimate and the actual population mean.

"Measure Precisely" [Hack #6] discusses how to use standard errors in the case of measurement. Calculating the *standard error of measurement* allows you to know how close your test score is to your typical level of performance. Just as measurement allows us to produce 95 percent confidence intervals around individual observed scores, statisticians routinely produce 95 percent confidence intervals around a wide variety of sample values.

Fortunately for anyone curious to know how far a statistical finding is from the hidden truth, every popular statistical procedure provides a standard error. After introducing the following basic concepts, this hack will explain how to apply the following standard errors:

- *Standard error of the mean* in descriptive statistics
- *Standard error of the proportion* in survey sampling
- *Standard error of the estimate* in regression

> The Central Limit Theorem **[Hack #2]** is a key tool for know-ing how wrong we are when we sample, because it provides the formula for calculating standard errors and suggests that all sample summary values are normally distributed.

There are three common ways that standard errors are used to verify the accuracy of results of statistical analyses. The particular tool you use depends on whether you want to know how close you are to correctly estimating:

- The mean score of a population on some variable (e.g., average salary of untenured college professors)
- The proportion of a population that have some characteristic (e.g., who will vote for my Uncle Frank as Chief Dogcatcher)
- Future performance (e.g., probable college GPA for your pet monkey, whom you have trained to take multiple-choice tests)

Mean Estimates

The precision of a sample mean as an estimate of a population mean is based on sample size. Here's the formula:

$$\text{Standard Error of the Mean } = \frac{\text{Standard Deviation}}{\sqrt{\text{Sample Size}}}$$

As the sample size increases, the closer the sample mean is to the true popu-lation mean. This makes sense if you think of sample size as the number of independent observations; the more looks you get at something, the more accurate your description will be.

> The *standard error of the mean* is the average distance of sample means from their population mean.

Proportion Estimates

When a sample of people is surveyed and the results are presented as some percentage or proportion (e.g., "72 percent of all sailors have knee trouble"), that percentage is some distance from the actual percentage you'd find if you surveyed the whole population. If the sample was selected randomly, the standard error of proportion indicates how close the sample percentage is to the population percentage.

The standard error of proportion is based on sample size and the size of the proportion. Here's the formula:

$$\text{Standard Error of the Proportion } = \sqrt{\frac{(\text{proportion})(1 - \text{proportion})}{\text{Sample Size}}}$$

Like the standard error of the mean, as the sample size increases, the size of the standard error of the proportion decreases. If you are mathematically oriented, you might notice that as the proportion moves away from .50, the smaller that number in the top part of the formula becomes.

When the calculations are made, then, the further the sample proportion is from .50, the smaller the standard error of the proportion. Another point of interest is that the top part of the formula is an indication of the amount of variability in the sample. (proportion)(1 − proportion) is the standard deviation for proportions squared.

> The standard error of the proportion is the average distance of sample proportions from the true proportion in the population.

Estimates of Future Performance

In regression analyses, scores on one or more variables are used to estimate scores on another variable [Hack #13]. However, that predicted score is unlikely to be exactly right.

Just as we can calculate how far an average sample mean is from a population mean or how far off our survey results are from theoretical population results, we can also say how far off, on average, our regression prediction will be from the actual score a person would get. Here's the formula:

$$\text{Standard Error of the Estimate } = \text{Standard Deviation}\sqrt{1 - \text{correlation}^2}$$

The standard deviation used in the equation is the standard deviation of the criterion variable, which is the one you are predicting. The correlation is the correlation between your predictor(s) and the criterion variable.

In the interest of accuracy (the point of this hack, after all), I should point out that the standard error of the estimate formula given earlier isn't quite correct. However, it does provide almost the same result as this more complex, but correct, equation:

$$SE_{estimate} = \text{Standard Deviation} \sqrt{(1 - r^2)\frac{\text{Sample Size} - 1}{\text{Sample Size} - 2}}$$

Notice with this formula that the larger the correlation, the smaller the standard error of the estimate. This makes sense, because if there is a lot of informational overlap between two variables, you can get a good sense of the score on one variable by looking at the other.

The standard error of the estimate is the average distance of the actual score from each predicted score.

Using Standard Errors

Here's how to use these tools to state with some confidence the range within which the truth lies. Because sampling errors are normally distributed, the standard error can be used just like a standard deviation to define specific proportions of scores under the normal curve.

For example, if we want to provide a range of values in which the population value falls 95 percent of the time, we can build a 95 percent confidence interval around our sample value. Based on the normal curve [Hack #23], 1.96 standard errors on either side of the sample value should provide a range of values that we can say with 95 percent certainty contains the population value.

Table 2-11 shows some examples of various standard errors and the use of sample data to produce these confidence intervals [Hack #6]. Notice how a larger sample size creates a sample estimate closer to the population value, and a larger sample size also points to a confidence interval that is more precise.

Table 2-11. Building 95 percent confidence intervals

Type of standard error	Standard deviation	Sample size	Sample value	Standard error	95 percent confidence interval
Standard error of the mean	15	30	100	2.74	94.63–105.37
Standard error of the mean	15	60	100	1.94	96.20–103.80

Table 2-11. *Building 95 percent confidence intervals (continued)*

Type of standard error	Standard deviation	Sample size	Sample value	Standard error	95 percent confidence interval
Standard error of the proportion	.25	30	.50	.09	.32–.68
Standard error of the proportion	.25	60	.50	.06	.38–.62
Standard error of the estimate	15	30	100	14.81	70.97–129.03
Standard error of the estimate	15	60	100	14.65	71.29–128.71

The "Sample value" column in Table 2-11 for the *standard error of the estimate* is an example of an estimated or predicted score on some variable. The calculations in the example assume a correlation of .25 between the predictor and the criterion.

Uncle Frank's Campaign for Dogcatcher

As the campaign manager for my Uncle Frank in his recent campaign for dogcatcher, I had an opportunity to use standard errors. Several weeks before the election, I surveyed 30 randomly chosen voters in the town of Tonganoxie, Kansas, where Frank lives. My survey found that 50 percent of respondents said they would vote for him. I warned Uncle Frank that the sample was so small that it was not a very precise reflection of the entire population of voters.

After referring to Table 2-11, I determined that if we had surveyed all the voters in town, the percentage saying they would vote for Frank might reasonably be anywhere between about 32 percent and 68 percent, though the most likely value was 50 percent. Of course, the optimist that is my uncle interpreted this as meaning he might have 68 percent of the vote and a huge lead. He spent the rest of his campaign chest on a giant victory party the night before the election. I, being the realist that I am and knowing my uncle's reputation around town, assumed the true outcome would be in the other direction. It was. That's okay, though. It was a great party.

Why It Works

We can trust the accuracy of standard errors if we accept the following assumptions and apply some common sense:

Sampling errors are normally distributed

This means that the size of these errors range in value in a way that matches the normal curve. This allows us to produce those persuasively precise confidence intervals.

Sampling errors are nonbiased

This means that sample values are equally likely to be greater or less than the population value. This is convenient because it means that across repeated studies, one can zero in on the true population value.

The formulas are constructed in such a way that if you have little or no information about the population, then the size of the *error* in your sample estimate is about the size of the standard deviation of the population.

Look what happens with the standard error of the mean or the standard error of the proportion when the sample size is 1, or what happens with the standard error of the estimate when the correlation is 0.00. Intuitively, a good formula for figuring the standard error size should produce smaller errors when more is known about the population.

Sample Fairly

#19   If you want to find something out about every single customer or employee in your business, you could talk to every single one of them. If you are concerned about the quality of the beer you serve at your bar, you could taste every one before serving. Or, to save time, money, and brain cells, "sample" efficiently instead.

Management thrives on knowing the characteristics of every widget produced, every transaction conducted, and every client helped. Of course, the whole set of all of these widgets, interactions, and people can never be brought together under one microscope and observed and evaluated. No specimen slide is big enough.

The same is true for those of us in social science—researchers interested in people simply cannot measure everybody. As much as we'd like to probe, shock, inject, hassle, embarrass, and generally bother everyone in the world, we just can't do it. We don't have the time, space, or money, and, frankly, no one really wants to get to know so many people.

The problem is, "How can you know about *everything*, without being able to look at everything?" As is the case with all hacks in this book, the solution is provided by statistics. There are scientifically sound ways to accurately describe any whole set of things by just looking at a small subset of those things.

Using Samples to Make Inferences

Inferential statistics allows us to generalize to a larger population, based on data from a smaller sample. For these generalizations to be valid, though, the sample has to represent the population fairly.

> A *population*, in the sense we use it here, is rarely the "population" of a country or city or planet in the way the term is used in social studies. In inferential statistics, a population is a description of the type of person or thing you're studying. Populations can be third-grade boys in Nebraska, nurses at Shawnee Mission Medical Center in Merriam, Kansas, South American giant otters, or books in the Library of Congress. The only rule is that a population is bigger than its corresponding sample.

A good *sample* represents a *population*. This means that the distribution of every important characteristic in a population must be distributed, proportionately, in the same way in the sample. Much of this hack is about how to construct a good sample, so let's look at a good sample.

Imagine a population of squares, diamonds and triangles, as shown in Figure 2-4.

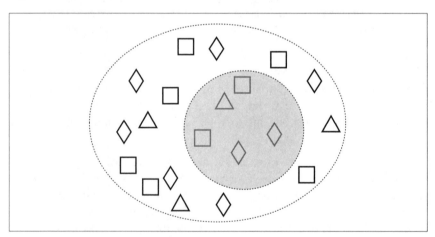

Figure 2-4. A sample within a population

A fair sample taken from a population of squares, diamonds, and triangles would contain those shapes in the same proportion as in the population. In our diagram, the outer oval represents a population, and the different shapes are distributed as 40 percent squares, 20 percent triangles, and 40 percent diamonds. The inner oval is the sample, which contains a subgroup of those elements in the population. The shapes in the sample are distributed in the exact proportions as in the population: 40 percent squares, 20 percent triangles, and 40 percent diamonds.

This sample is fair. It represents the population well, at least in terms of the characteristic of *shape*. When sampling people or things, samples typically represent a variety of traits. People and things are not entirely *triangles* or *squares*, so a sample of people is representative when its mean level of traits matches well with the population levels. Each person will have some level of all the characteristics, and won't be *entirely* one trait, unlike our shape example. (Though my Uncle Frank *is* pretty much entirely square, according to my Aunt Heloise.)

> The person asking the question gets to pick the population he is interested in, but he is then accurate when generalizing to that population only, not any other.

If you knew that the sampling methods used to produce this sample (the elements in the inner oval) were correct, you could infer something about the population by just looking at the sample. The procedure is simple and intuitive:

1. Observe the sample. For example, 20 percent of the sample is triangles.
2. Infer to the population. I bet 20 percent of the population is triangles.

Instead of abstract triangles in a theoretical population, imagine you are interested in checking the quality of the beer you sell in your bar. To get an idea of the beer population, construct a good sample of the beers you sell and taste each of them:

1. Observe the sample. For example, 20 percent of the beers have just a hint of a possum aftertaste.
2. Infer to the population. I bet 20 percent of all the beers you sell have just a hint of a possum aftertaste. You might consider cleaning your beer tap.

Inference is pretty easy to do, but it works well only when the sample is good. Constructing a good sample is the key.

Constructing the Best Random Sample

A good sample represents the population. Representative sampling begins with defining the universe, or, in other words, the population of things from which a researcher wishes to sample. There are a variety of ways to conceptualize these elements and various levels of grouping that are explicitly or implicitly identified when choosing a population and selecting a sample.

You have to know about these ways of organizing your population; otherwise, you cannot create a good sample:

General universe
> Abstract population to which a researcher hopes to generalize his findings. For example, I might want to say something about all *comic book collectors*.

Working universe
> Concrete population that allows for sampling to occur. I can't really be sure I have located or counted all comic book collectors, but I could operationalize that population by defining it as all the *subscribers* to the *Comics Buyer's Guide*, a monthly magazine that most serious collectors read. This working population is not exactly the same as the general universe, but it should be almost as large and will capture most of the abstract population of interest.

Sampling unit
> Element that defines the population. In our example, a single subscriber to the monthly magazine would be a sampling unit.

Sampling frame
> List, real or imagined, of sampling units in a population. In our example, this would be the literal list of subscribers that I might be able to purchase from the magazine.

> An observation that is likely true about the people and things that were not part of your sample is said to be *generalizable*. If a sample does not represent a population, the sample is *biased* (a bad sample).

The best sampling strategy, without question, is to sample *randomly* from a valid sampling frame. Random selection will do the best job of creating a sample that represents all the traits of interest in the population. The real power of random selection, though, is that you are also representing all sorts of variables you haven't even considered that might otherwise have an impact on your observations.

Technically, the term *random* describes a sampling process that gives every member of a population an equal and independent chance of being selected. *Equal* means that every sampling unit in the sampling frame has as good a chance as anyone else. *Independent* means that a person's or thing's chances of being selected are unrelated to whether any other particular person or thing has been selected.

So, suppose a selection process calls customers on a client list to ask for participation but stops trying to contact people if they aren't home or in the office during the first attempt—this does not give all possible participants an equal chance of being selected. People who aren't easily available are less likely to be chosen, and if people are not solicited to participate when someone in their office has already been chosen, each member of the population does not have an independent chance of being chosen.

Random sampling can be done by numbering all names on the sampling frame list and using some method of choosing a random number to pick each participant.

Sampling Strategies for the Real World

In the real world, it is often difficult or impossible to sample randomly. Here are some sampling strategies that aren't quite as good as random sampling, but are more realistic outside of some imaginary scientific laboratory:

Convenience sampling
> The sample is chosen based on accessibility. This is sometimes called *haphazard sampling*. Head down to the local mall and ask the first 10 people you see how they feel about your company's widgets, and you have engaged in convenience sampling.

Systematic sampling
> Units are chosen from the sampling frame at equal intervals. For example, you might take every 10th person from a long list. As long as the order of names on the list is unrelated to whatever you are trying to determine, this might do as good a job of representing the population as true random selection. Statistical theorists and practitioners actually have academic debates over this issue.

Stratified sampling
> The sampling frame is divided into meaningful subgroups, and units are randomly chosen from each subgroup. This could result in greater representativeness than even random sampling if the characteristics that define the subgroups are important to the question you are asking.

Cluster sampling
> Groups of units are randomly chosen, and all units in those groups are sampled. For example, you might choose a publishing company at random and then interview every employee about how to succeed in publishing.

Judgment sampling
> The sample is chosen based on your expert judgment as to whether the sample will represent the population. You might choose to talk to only your best customers, because they know the most about your widgets.

Choosing a Sample Size

If you are able to construct a *good sample*, as we have defined it, even a small sample can be effective. As with chocolate chip cookies, though, bigger is better. The larger the sample, the more representative of the population it is. Consequently, the observations are more generalizable and you can better trust their accuracy.

Also, if there is some interesting relationship between variables in your observations, you are more likely to find that relationship and be sure that it did not occur by chance when you have observed many elements in your sample than when you have looked at just a few.

Finally, if you do have some social science purpose for your sampling, there are certain technical statistical characteristics that must be met to perform certain analyses. These standards are easier to meet in larger samples—such as, say, samples consisting of 30 or more widgets.

See Also

- "Find Out Just How Wrong You Really Are" [Hack #18] shows how to determine error size in inferential statistics.

HACK #20   Sample with a Touch of Scotch

When statisticians choose samples of people from populations, they are really sampling from continuous distributions of variables. Sampling is sometimes easier to understand, though, by treating your variables as discrete objects, not continuous scores.

The most powerful statistical procedures use scores at the interval level of measurement or higher [Hack #7]. To sample scores from a population, social science researchers usually choose *people*, though, not scores. The people are then measured, which results in a sample of scores. So far, so good.

When discussing the sampling process, however, smart researchers some-times sound not-so-smart when they refer to their sampling strategy. For example, if a researcher is interested in measuring the effects of some treat-ment on a *continuous* variable such as *happiness*, he might say (and think), "OK, first I need to get a sample full of happy and unhappy people." He, at least for the moment of the thought, is treating happiness as if it were a *dichotomous* variable.

> *Dichotomous* is statistics jargon meaning "having only two val-ues." For example, biological sex is a dichotomous variable.

He is referring to people as if they are either completely happy or entirely unhappy. In reality, of course, he thinks there is a large range of happiness scores that describe people, which is why he is using statistics that make the assumption of interval measurement.

He refers to his participants as *either/or* because doing so makes it easier for him to picture the representativeness of his sampling. It's a smart strategy, because by thinking of samples as representing big, discrete categories instead of more precise, continuous values, this sometimes makes questions about sampling easier to answer and justify.

A Sampling Problem

Here's a brainteaser that centers on a sampling question. A drunk, untenured statistician (I've met a few) is mixing drinks at a party. He is making a Scotch and soda for his department chair. The chair demands a drink with some exact proportion of Scotch to water (it doesn't matter what the specific request is; our hero never makes it that far).

The statistician starts with two glasses of the same size. One glass (the first glass) has two ounces of Scotch in it; the other (the second glass) has two ounces of water in it. He starts by pouring an ounce of water from the water glass into the Scotch. He apparently already screwed up, because he changes his mind and pours an ounce of the new mixture (three ounces of Scotch and water mixed up) back into the water glass. Both glasses now have two ounces of liquid in them, but the liquid in each glass is some mix of water and Scotch.

Nervously, the statistician attempts to start all over, but his department chair stops him. She says:

> I have a proposition for you. We can't possibly know the exact proportion of Scotch and water in each glass right now, because we can't know how mixed up everything is. But if you can answer the following question correctly, I'll write a strong letter of support to your tenure committee. If not, well, I'm sure someone with your qualifications should have no trouble finding work in the hotel/motel or food service industry. Here's the question: right now, does the first glass have more water in it, or does the second glass have more Scotch in it?

Think of the question as a sampling issue. Does the first sample, the liquid in the first glass, have more water in it, or does the second sample, the liquid in the second glass, have more Scotch in it? Because both Scotch and water are made up of really small particles, it is difficult to picture how much of each liquid is represented in each sample. Even proportionately, we can't be sure how many water particles (or sampled scores that equal "water") are mixed into the sample of "Scotch" scores, because who knows how much water drifted down into the bottom of the first glass and would have remained there as the top part of the liquid near the surface was poured back into the second glass. An intuitive answer is called for. Unfortunately, it is wrong.

The intuitive answer typically generated by smart people is that the first glass, the Scotch glass, has more water in it than the water glass has Scotch in it. This makes sense because pure water was poured into the Scotch, while some mix of water and Scotch was poured back into the water glass. Amazingly, this clever thinking leads us astray. The correct answer is that the proportions are equal! There is the same amount of water in the Scotch glass as there is Scotch in the water glass.

Using Metaphor to Solve the Problem

The solution to the sampling problem is clearer if we imagine that our variables are not tiny particles, but instead are large categories, such as blue and white marbles. Instead of a glass of Scotch, imagine a glass of 100 blue marbles. Instead of a glass of water, imagine a glass of 100 white marbles.

The glasses are big, so the marbles can get mixed together well. Think large glass fishbowls. This is necessary to ensure that random selection is possible, as was likely with the mixed-up liquids. Keep your eye on the marbles through each step of the mixing.

Our hero takes 50 white marbles from the second glass and mixes them into the first glass. The distribution of the two variables is now:

Sample 1
　100 blue marbles, 50 white marbles

Sample 2
　50 white marbles

Now, he (randomly, remember, to simulate the mixed liquids) takes any 50 marbles from the first glass and mixes them back into the second glass. Let's imagine a variety of possibilities.

If by chance he selects all the white marbles, they go back into the second glass and the distribution is now:

Sample 1
　100 blue marbles

Sample 2
　100 white marbles

If by chance he selects no white marbles and puts 50 blue marbles into the second glass, the distribution is:

Sample 1
> 50 blue marbles, 50 white marbles

Sample 2
> 50 white marbles, 50 blue marbles

Now imagine a more likely scenario: some of the marbles he randomly draws are white and some are blue. For example, he could draw out 10 white marbles and 40 blue marbles and place them in the second glass. In that case, the new distribution is:

Sample 1
> 60 blue marbles, 40 white marbles

Sample 2
> 60 white marbles, 40 blue marbles

Try this with any mix of marbles you wish, but remember you have to draw out a total of 50 marbles (to duplicate the one ounce, or half, of the water originally mixed up).

Notice that any mixture you try results in 100 marbles in each glass at the end. Also, most importantly, notice that the ratio of blue to white marbles in the first glass at the end is always equal to the ratio of white to blue marbles in the second glass. Any blue marble that is not in the second glass must be in the first glass, and any white marble that is not in the first glass must be in the second glass.

The same is true for Scotch and water. The correct answer is that the proportions will be equal, no matter how they were originally mixed up.

Where Else It Works

Real-life polling companies, who make their living and stake their reputations on the accuracy of election predictions, are also primarily concerned with the proportion of samples who are in each of several crucial categories. If people have just voted and there are two candidates, anyone who did not vote for candidate A voted for candidate B. Their absence in one category guarantees their presence in the other. Reporting predictions as percentages creates the potential for greater accuracy. It also allows for greater error, as a voter predicted to be in category A who ends up in category B has therefore produced error in both categories.

When statistical social science researchers want to be convinced that their sample is representative of its population, their primary concern is always the proportions of characteristics in their sample, not the number of people with those characteristics. What matters most is that the proportions of each score for the key research variables are the same in both samples and their populations.

Choose the Honest Average

Data-driven decisions, such as whether you can afford to buy a house in a new town or who the core market is for your business, often rely on the "average" as the best description for a large set of data. The problem is that there are three completely different values that can be labeled as the "average," and the different averages often result in different decisions. Make your decisions using the correct average.

When most people hear a statement like "the average price for a house in this town is $290,000" (which might sound low, high, or just right, depending on where you call home), they imagine that this figure was determined by adding up all of the sales prices from all of the houses in the town, and then dividing that sum by the number of houses. But statisticians know there is more than one way to determine the "average," and sometimes one kind is better than another.

Whether that $290,000 really represents the typical housing price depends on whether the average is actually the *mean*, *median*, or *mode*. It also depends on the shape of the distribution of all the numbers that are averaged. Wise folks will make sure they are making their decisions using the best summary value. Here's when to trust each type of average.

Measures of Central Tendency

The purpose of determining an average for a set of values—whether those values are house prices, grades from a final exam, or the number of students in a yoga class—is to efficiently communicate the *central tendency* for those values. It's true that, most of the time, central tendency is determined by adding up all of the values in a distribution, and then dividing the sum by the number of values. Statisticians don't call this the *average*, though; they call it the *mean*. So, why not always use the mean to determine central tendency? Because in some situations, the mean doesn't represent any of the *actual* values!

Consider the opening example about the average price of a house. Let's say you collect data for 300 houses in a town and want to determine the average sales price in that sample. Generally speaking, the mean is not a very good indicator of central tendency for house prices. Figure 2-5 illustrates why.

The mean is not a very honest average in this situation, because the distribution of sales prices is skewed by a few outlying values that are very large. Of the 300 houses sampled, 231 of them were sold for prices in between $100,000 and $600,000. The remaining 69 houses sold for prices above $600,000, with 56 of those above a million dollars. The mean is heavily

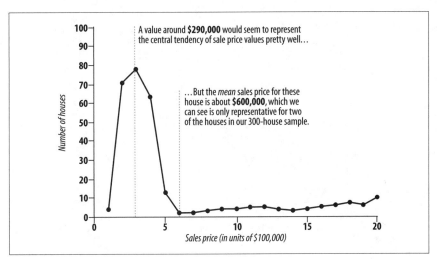

Figure 2-5. Mean as a misleading average

influenced by these outlying values and therefore is not very representative of *any* house in the sample.

Means don't work well as averages for most money variables. The average income reported as a mean is much higher than what most people earn. There are always a few Bill Gates and J.K. Rowling types who pull the mean way up.

So, what's the "honest average" for these types of values? Instead of reporting the mean, with distributions like the one in Figure 2-5, honest statisticians generally prefer the *median*. The median is that value in a distribution at the 50th percentile, such that half of all values are below it and the other half are above it (just like, on a highway, the *median* divides the road in half). The median for this distribution of data is just under $290,000, and thus works very well as a measure of central tendency.

Choosing the Middle Ground

The median works well in these instances because it is much less sensitive to outlying values than the mean, and thus is preferred whenever a distribution is skewed in one direction or another. The median is therefore also the most "honest" measure of central tendency when the distribution is skewed by a few outlying values that are much *smaller* than the rest, as in Figure 2-6, a fictional set of 50 students' exam scores.

Figure 2-6 shows another type of data in which a mean might lead to a wrong conclusion. Relying on the median here would result in a more accurate interpretation of class performance.

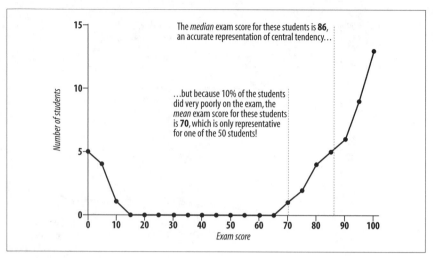

Figure 2-6. Median as the honest measure of central tendency

Where It Doesn't Work

Not even the median will always be honest, though. Consider the following scenario. Say you're a yoga instructor, and half of the students in your class are between 25 and 35 years old, and the other half are between 50 and 60. How would you describe the average age of your students?

The problem in situations like these is that neither the mean nor the median will adequately describe the group of individuals. What to do? The most honest choice for an average in this situation is to report the *mode*, which is simply the most frequently occurring value in a sample of data, as shown in the example in Figure 2-7.

In this case, there are two modes: one at 30 years old and the other at 54 years old. Reporting both of these values is the best way to choose the honest average. The mean and median both mislead for these sorts of data.

How to Choose the Honest Average

So, when *is* the mean the honest average? Basically, the mean is the best choice when there is only one mode and the distribution is symmetric, which means that there is no obvious skew in either direction. If your yoga class were attended by your 25- to 35-year-old students only, the mean would be the honest average.

When all is said and done, how do you choose the most appropriate average? Following these three simple rules will keep you honest if you are

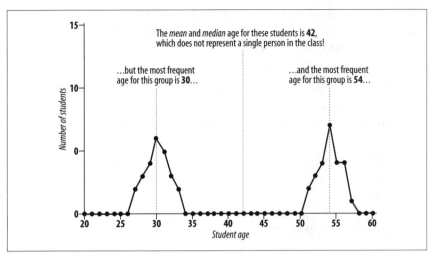

Figure 2-7. Mode as the honest average

reporting summaries, and will help you make informed choices if you are the one making decisions based on the data:

- Choose the *mode* if there are two or more "trends" in the data (i.e., two or more areas of high-frequency values), and report one mode for each trend.
- Choose the *median* if the distribution is skewed (i.e., a small number of outliers is heavily influencing the mean).
- Choose the *mean* if the distribution is fairly symmetric with one mode.

It is interesting to note that in most cases, the mean, the median, and the mode will all be fairly close to equal. So why bother with the mean? The mean remains as the most common way to report the average because it is most likely to be replicated if we were to take another sample of data and look for the central tendency. Medians and modes tend to be a lot more variable, but the mean stays nice and stable.

—*William Skorupski*

HACK #22 Avoid the Axis of Evil

Graphs are powerful tools to represent quantities, relationships, and the results of research studies. But in the wrong hands, they can be made to deceive. Choose your destiny, young Luke (or, if you are under the age of 25, "young Anakin"), and avoid the dark side.

There was a time when only scientists, engineers, and mathematicians ever saw a graph. With the advent of more and more news outlets aimed at the general public, visual representations of numeric information have become more and more common. Just think of yesterday's issue of *USA Today*—it contained at least a dozen graphs.

In business conferences, graphs are used frequently to communicate information and demonstrate success (or failure). If the creator of a graph isn't careful, though, choices that might seem arbitrary will affect the interpretation of the information. Without changing the data, you can change the meaning.

So, if you want to avoid manipulating your audience when you create a graph, or if you just want to be able to spot a misleading (whether intentional or not) chart, then use this hack to help you create and interpret graphs effectively.

Choosing the Honest Graph

To understand correct and incorrect graphing options, we first have to cover some graphing basics. There are various pieces to a graph, and the manipulation of those pieces can lead or mislead.

Typical graphs have two *axes*, because they describe two different variables. Axes are the lines along the bottom, called the *X-axis*, and along the side, called the *Y-axis*.

> You can remember that the vertical axis is called the Y-axis because the cute little letter Y is reaching its cute little hands up, vertically, toward the sky. Get it? (Welcome to the creative world of statistics education.)

The sort of graph that is appropriate (and nondeceptive) for showing the variables you have measured depends on the level of measurement of your variables [Hack #7]. You can choose from three common types of graphs, and only one will be the right one for your variables:

Bar chart
    In Figure 2-8, the X-axis represents categories or groups, such as males and females. The Y-axis is continuous: the taller the bars, the higher the value on variable Y.

Histogram
    In Figure 2-9, the X-axis represents continuous values. A histogram is often used when the X-axis represents common categories that reflect an underlying continuous variable, such as months of the year or some other distinctive set of groupings that can be placed in a meaningful order. These look like bar charts, except that the bars are pushed together with no spaces between them.

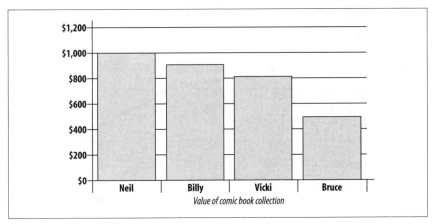

Figure 2-8. Bar chart

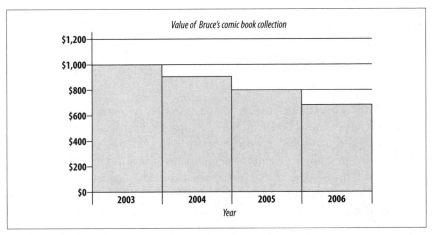

Figure 2-9. Histogram

Line chart

In Figure 2-10, both the X- and Y-axis are continuous variables; in this example, they're time and value. The higher the line at any point, the greater the quantity as represented by the Y-axis.

To pick the right kind of graph (i.e., the one with the format that is the least deceptive and the most intuitive), identify the types of X variable you are using (notice that Y is *continuous* in all of these formats):

- If X represents different categories and Y is continuous, use a bar chart.
- If X can be conceived of as categories, but there is also some meaningful order among them and Y is continuous, use a histogram.
- If X and Y are both continuous, use a line chart.

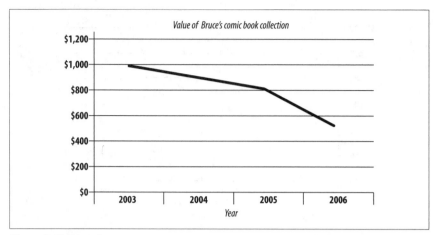

Figure 2-10. Line chart

Graphic Violence

A common error in graphing, either intentional or not, has to do with set-ting the scale for the X-axis. Here's why this is a problem and how you can avoid it.

Graphs with two variables invite comparisons across categories or time or across different values of one variable. Pictures are worth a thousand words, as they say, and a graph can be very persuasive evidence. Anytime lines or bars are used to compare values, the comparison is accurate only when the height of the line or the length of the bar is judged against some standard minimum value. That minimum value is often zero. If the graph is not cali-brated to some reasonable base value, small differences look huge.

Compare the two graphs shown in Figure 2-11, for example. Both convey exactly the same data, and yet your interpretation of each might be wildly different. The histogram in the top left reflects performance of the U.S. stock market over the last five days. Notice a rather frightening-looking drop on day five. No doubt, earth-shaking news hit near the end of day 4. You might also notice that the Y-axis (the Dow Jones Index) does not begin at zero; it begins at 9,900, a value that is low enough to contain the top of all five bars, but that is otherwise not meaningful.

Look more closely at the second histogram in Figure 2-11, on the bottom right. Both charts present the same data, but the second graph uses zero as the starting point. The interpretation of the data as presented in this graph shows very little fluctuation across the last five days, and the frightening drop at day 5 is barely a hiccup.

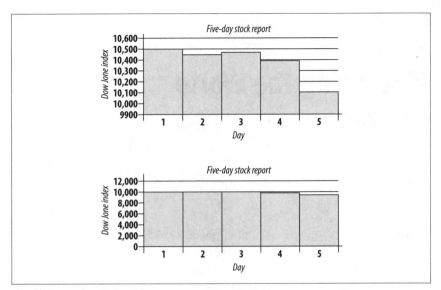

Figure 2-11. The power of the Y-axis

Which display is the correct one? Both reflect a drop of 2.8 percent in stock market value from day 4 to day 5. It really depends on the intent of the graph constructor and the intended audience. When number counts are involved, or money, the most meaningful and fairest starting point is usually nothing. Many newspapers provide daily stock information in the format as shown in the first histogram. They believe their readers are interested in small changes, so they set a Y-axis starting value that is as high as possible but low enough to contain all data points on the X-axis.

After all, to an avid investor who changes her portfolio often and buys and sells frequently, a drop of 2.8 percent is serious business. A graph designed to make small changes look serious might be the most valid for that reader. If an investor is one of those "in it for the long haul" types, a relatively small change is meaningless, however.

To get the most meaning out of graphs like these, always check the bottom value on the Y-axis. This way, you can get a sense of the real differences on the X-axis as you crawl from bar to bar. If you are making graphs like these, think about the most honest way to present the information. You want to inform, not deceive (probably).

See Also

- The book that first pointed out to the general public how charts can deceive, especially in advertising, was How to Lie With Statistics. Huff, D. (1954). New York: Norton and Company.

CHAPTER THREE

Measuring the World
Hacks 23–34

There is great value in understanding phenomena by hanging a quantity on it. Though sometimes a something important is lost in the translation from idea to number, creating scores to represent whatever we are interested in does allow for a level of precision in understanding, and it also allows for comparison. These hacks all involve measurement and interpretation of scores.

A whole family of hacks relies on the normal distribution [Hack #23] and its presence everywhere we look. With the normal curve, you can tell where you stand compared to everyone else [Hack #24], know how you are likely to perform on a test before you even take it [Hack #25], and understand your test results at a deeper level [Hacks #26 and #27].

Speaking of testing, you'll learn how to produce a good set of questions [Hack #28] and make a quality test [Hacks #31 and #32]. You can identify bad items, worthless questions, and do well on a test without knowing the answers [Hack #29]. You can also improve your test performance without cracking a single book [Hack #30].

Finally, by learning a couple of solid measurement principles, you can determine the lifespan of an era, person, or business [Hack #33] and also learn how to use medical information [Hack #34] to maybe increase your own lifespan.

Measure by measure, here is a whole chapter full of measurement hacks.

See the Shape of Everything

Almost everything in the natural world is distributed in the same way. As long as you can measure the thing, whatever it is, and scores are allowed to vary, it has a well-defined "normal distribution." If you know the specifics about the shape of this normal curve, you can make very accurate predictions about performance.

There are a few miracles in the world of statistics. There are at least three tools—three discoveries—that are so cool and magical that once students of statistics learn about them and begin to comprehend their beauty, they frequently explode.

Well, maybe I am exaggerating a bit, but here are three dandy tools for understanding the world:

- The correlation coefficient **[Hack #11]**
- The Central Limit Theorem **[Hack #2]**
- The normal curve

Since we've discussed the uses of the first two miracles in other hacks, let's spend our time now getting to know the shape and uses of the third: the *normal curve*. I am pleased to present the normal curve, the normal distribution, the bell-shaped curve, the whole world, as shown in Figure 3-1.

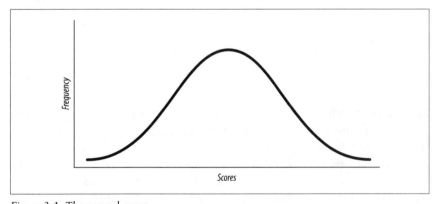

Figure 3-1. *The normal curve*

Applying Areas Under the Normal Curve

Statisticians have defined the normal curve very specifically. Using both calculus and hundreds of years of real-world data collection, the two methods have reached the same set of conclusions about the exact shape of the normal distribution. Figure 3-2 shows the important characteristics of the normal curve. The mean is in the middle, and there is room for fewer and fewer scores as you move away from that center.

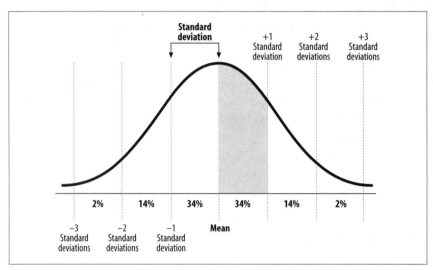

Figure 3-2. Areas under the normal curve

Though the normal curve is theoretically infinitely wide, three standard deviations on either side of the mean is usually enough to contain all the scores.

> A distribution's *standard deviation* is the average distance of each score from the mean [Hack #2].

Predicting test performance. Recall the claim I made earlier that anything you measure will distribute itself as a normal curve. By implication, then, anything we measure will have most of the scores close to the mean and only a few scores far from the mean. Measure enough people and you will get the occasional extreme score very far from the mean, but scores far from the mean will be rare. The expected proportion of people getting any particular score gets smaller as that score moves away from the mean.

That next test you take? I don't know the test or anything about you, but I am willing to wager that you will get a score close to the mean. I predict your score will be average. You might get above average or below average, but the normal curve tells me that you will likely be pretty close to the mean.

To make these sorts of predictions, and to be pretty confident about their accuracy, you can use the known normal curve's dimensions to estimate the percentage of scores that will fall between any two points on the X-axis (the bottom, horizontal part of the graph). The percentage of scores between pairs of standard deviation points on the scale are shown in Figure 3-2. The percentages add up to 100 percent, but that is because of rounding. Remember that some scores, though just a few, will be further than three standard deviations away from the mean.

Here are some key facts about the curve that you can use to predict performance:

- About 34 percent of scores fall between the mean and one standard deviation above the mean. See the shaded section in Figure 3-2? If you took some ink and colored in the entire space beneath the normal curve, you would use 34 percent of the ink on this section.
- About 34 percent of scores fall between the mean and one standard deviation below the mean.
- About 14 percent of scores fall between one and two standard deviations above the mean.
- About 2 percent of scores fall between two and three standard deviations below the mean.

You can also combine the percentages to make other statements such as:

- About 68 percent of all scores will be within one standard deviation of the mean.
- About 50 percent of scores will be below the mean.

You can use these known percentages to make predictions and statements of probability. We can speak of the normal curve as either the *percentage of scores* that fall under given areas on the curve or the *likelihood that any given test taker* will fall under given areas:

- There is a 2 percent chance that you will score more than two standard deviations above the mean on your next test.
- There is only a 16 percent chance that this applicant will score lower than one standard deviation below the mean on our job skills test.

Setting standards. Policy makers rely on the assumption that ability is normally distributed when they establish levels of performance. They choose levels of performance that will guarantee them a certain percentage of qualifying people. The normal distribution is an invaluable tool for setting policy for admissions or services if one wants to magically know ahead of time how many people will qualify.

For example, a college with high academic standards might require scores on an ability test that are at least one standard deviation above the mean. This way, they ensure themselves of accepting only the top 16 percent in ability.

Likewise, special education policy in the United States establishes certain cut scores for students on tests that qualify them for special education status (and, thus, federal and state funding). Cut scores are specific scores that a person must score above (or below). If policy makers have the budget to pay for special programming and staff for only, say, two percent of all children, they set the cut score at two standard deviations below the mean. Faith in the normal curve allows them to calculate the number of children who will need funding.

Appreciating the Beauty of the Normal Curve

To appreciate the wonder of the normal distribution, you can always build your own. Imagine you measured something (such as attitude, knowledge, height, or speed). You have some scoring system in which scores are allowed to vary (such as scores on an attitude survey, or SAT scores, or inches, or miles per hour). You have lots of scores because you measured lots of people, buildings, or sparrows. Now, plot these scores on a graph such that the X-axis represents the actual score value from lowest to highest, left to right (or the other direction if you'd like). The Y-axis (the vertical left side part) should represent the relative frequency of each value in your group of scores.

On such a chart, the height of the line or dot represents the relative proportion of scores that were at any particular value. Notice on the normal curve that the highest points are in the middle and the lowest points are on the ends. The middle score is the average score and the most popular score. On the normal curve, the median is equal to the mean, which is equal to the mode [Hack #21].

Notice also that the normal curve is symmetrical: you could fold it in half and one side would perfectly cover the other. The other characteristic of the normal curve that is important to know is that it goes on forever. It is a theoretical curve, so the two ends of the curve will never touch the baseline.

The normal curve is the common truth that connects all of nature. It is perfectly balanced. It is forever. It is eternal. It also kind of looks like a dinosaur, which is cool.

Produce Percentiles

HACK #24

A simple but powerful way of understanding test performance is through the use of percentile ranks. Here's how to take a raw score with little explanatory value and transform it into something much more informative and useful.

In school, teachers (or counselors, or whoever reported standardized test results) might have reported results to you without ever telling you your score. Instead, you probably saw a number that looked like a percentage and was described as telling you how you (or your child) compared to others who took the test. This type of score is called a *percentile rank*.

If you have been shown a percentile rank that represents your test performance, it won't be useful unless you know what it means. On the flip side, if you have to explain someone's test performance and you show the test taker a raw score only, you aren't really being very helpful. Being able to build or interpret percentile ranks is a useful skill for both sides of the testing game.

Norm-referenced scoring [Hack #26] is an approach to making test scores more informative by comparing scores to each other. The norm-referenced score you see most often in the real world is the percentile rank. The percentile rank is defined as "the percentage of scores in a distribution that are less than a given score of interest." For example, if you get 15 items correct out of 20 on a quiz and half the class got fewer correct than you, your percentile rank is 50.

Producing and Reporting Percentile Ranks

If you are a classroom teacher, human resources manager, or anyone who has to report test results to others, being able to report a percentile rank instead of a raw score will help test takers understand how well they performed and also help decision makers understand the consequences of setting various standards of performance.

Organize your data. Producing percentiles begins with organizing all your test scores. For a small data set, it is fairly simple to build a *frequency table*, which answers all sorts of questions in addition to providing percentile ranks. Here is a sample distribution for 30 scores on a classroom test (arranged from lowest to highest) in which 100 points was the highest possible score:

> 59, 65, 72, 75, 75, 75, 80, 83, 83, 85, 85, 85, 85, 85, 85, 86, 86, 86, 86, 88, 88, 88, 90, 90, 90, 90, 90, 92, 94, 97

Compute frequencies and percentages. For efficiency's sake, this data can be displayed and the frequency of each score can be computed, as shown in Table 3-1.

Table 3-1. Cumulative frequency for a classroom test

Score	Frequency	Cumulative frequency	Percentage	Cumulative percentage
59	1	1	3.33 percent	3.33 percent
65	1	2	3.33 percent	6.67 percent
72	1	3	3.33 percent	10.00 percent
75	3	6	10.00 percent	20.00 percent
80	1	7	3.33 percent	23.33 percent
83	2	9	6.67 percent	30.00 percent
85	6	15	20.00 percent	50.00 percent
86	4	19	13.33 percent	63.33 percent
88	3	22	10.00 percent	73.33 percent
90	5	27	16.67 percent	90.00 percent
92	1	28	3.33 percent	93.33 percent
94	1	29	3.33 percent	96.67 percent
97	1	30	3.33 percent	100.00 percent

Table 3-1 shows each score that someone actually got, how many people got that score, the total number of people getting a given score or lower, the percentage of all people getting each score, and the total percentage of people getting a given score or lower. The *cumulative* columns always report the total number of people (or scores) in the distribution (30 in our example) and the total percentage of people (always 100 percent).

Determine percentile ranks. To determine the percentile rank for any score in the distribution, use the "Cumulative percentage" column. Find the score of interest and look at the cumulative percentage in the row *just above* that score's row. For instance, for a score of 94, the percentile rank is 93.33 or about the 93rd percentile. For a score of 86, the percentile rank is 50.

> If you review a dozen statistics or measurement textbooks, you'll find that there are actually two different and competing definitions for a percentile rank. I prefer "the percentage of scores in a distribution that are less than a given score of interest," but some books give "the percentage of scores in a distribution that are *equal to* or less than a given score of interest." Both definitions are reasonable and percentile ranks can be calculated either way using a frequency table. Under the first use of the term, there can be no 100th percentile. Under the second, there can be no 0th percentile. Pick the definition you prefer and go with it, but always share your definition along with your results.

Interpreting the Percentile Rank

Imagine that you are sitting down with your guidance counselor and have been told that your percentile rank is 93. So, what does this mean? Well, the most direct interpretation is that 93 percent of all people who took the test scored less than you did. It is also correct to say that 7 percent of people scored equal to you or higher. We can also think of percentile ranks as saying how far the score is from normal. The mean percentile rank is usually around the 50th percentile and will be exactly that if scores are normally distributed, as they (ahem) *normally* are. So, we could also say that the 93rd percentile is pretty far above average.

Don't make the mistake that many otherwise savvy stat-hackers sometimes make. Earlier in this hack, we used an example of a test score in which you got 15 items correct out of 20 on a quiz and half the class got fewer correct than you. Your percentile rank in that example was 50. Notice that in that example, your percent correct is 75 percent (15/20), but your percentile rank is 50. Don't confuse the two! Knowing your percentile rank does not tell you how many questions you got right.

Where It Doesn't Work

Remember that a percentile rank is useful only when you're looking for a norm-referenced interpretation. If you want to know whether you have mastered a key set of skills, it does not help to know what percentage of people have mastered more or less of those skills. To know where you are compared to some set of standards, not compared to other people, you want a criterion-referenced score [Hack #26]. A *percent correct* type of score is more meaningful for you in this case than a *percentile rank*.

See Also

- If you assume that your scores are normally distributed, or at least drawn from a population that is normally distributed, you can just convert any standardized score directly to the percentile rank, using information about the areas under the normal curve [Hack #25].

Predict the Future with the Normal Curve
HACK #25

Because almost anything we measure in the natural world has a known distributional shape, the "normal curve," we can use the precise details of that distribution to predict the future and answer all sorts of probability questions.

A variety of hacks in this book capitalize on statisticians' close personal relationship with the *normal curve*. "See the Shape of Everything" [Hack #23] shows how to use the normal curve to predict test performance in a general way. We can do better than that, though.

So much is known about the exact shape of this mystical curve that we can make exact predictions about the probability that scores in a certain range will be obtained. There are many other types of questions that can be asked related to test performance, and statistics can help us to answer these sorts of questions before we ever take the test!

For example:

- What are the chances that you will score between any two given scores?
- How many people will score between those two scores?
- What are the chances that you will pass your next test?
- Will you get accepted into Harvard?
- What percent of students in the U.S. will qualify as National Merit Scholars?
- What are the chances that my Uncle Frank could pass the Mensa qualifying exam?

For these types of questions, a precise tool is needed. This hack provides that tool: a *table of areas under the normal curve*.

The Table of Areas Under the Normal Curve

The normal curve is defined by the mean and standard deviation of a distribution, and the shape of the curve is always the same, regardless of what we measure, as long as the scoring system allows scores to vary. The proportions of scores falling within various areas beneath the curve, such as the space between certain standard deviations and distances from the mean, have been specified.

This hack relies on a complicated-looking table, but it is so full of useful information that it will quickly become a primary tool in your hacker's toolbox. Without further ado, take a deep breath and look at Table 3-2.

Table 3-2. Areas under the normal curve

z score	Proportion of scores between the mean and z	Proportion of scores in the larger area	Proportion of scores in the smaller area
.00	.00	.50	.50
.12	.05	.55	.45
.25	.10	.60	.40
.39	.15	.65	.35
.52	.20	.70	.30
.67	.25	.75	.25

Table 3-2. Areas under the normal curve (continued)

z score	Proportion of scores between the mean and z	Proportion of scores in the larger area	Proportion of scores in the smaller area
.84	.30	.80	.20
1.04	.35	.85	.15
1.28	.40	.90	.10
1.65	.45	.95	.05
1.96	.475	.975	.025
4.00	.50	1.00	.00

Deciphering the Table

Before we use this nifty tool, we need to take a second deep breath and get the lay of the land. I have simplified the information on this table in a couple of ways. First, I have listed only a few of the values that could be computed. Indeed, many tables in statistical books have every value between a z of .00 and a z of 4.00, increasing at the rate of .01. That's a lot of information that could be presented, so I have chosen to show only a glimpse of the most commonly needed values, including the z scores necessary for 90 percent confidence (1.65) and 95 percent confidence intervals (1.96); see "Measure Precisely" [Hack #6] for more on confidence intervals.

I have also rounded the proportions to two decimal places. Finally, I used the symbol z in the table to indicate the distance from the mean in standard deviations. You can learn more about z scores in "Give Raw Scores a Makeover" [Hack #26].

After understanding the simplifications made to the table, the first step toward using it to make probability predictions about performance or answer statistical questions is to understand the four columns.

The z column
> Picture the normal curve [Hack #23]. If you are interested in some score that could fall along the bottom horizontal line, it is some distance from the mean. It could be greater than the mean score or less than it. The distance to the mean expressed in standard deviations is the z score. A z score of 1.04 describes a score that is a little more than one standard deviation away from the mean. Because the normal curve is symmetrical, we don't bother to note whether the distance is negative or positive, so all of these z scores are shown as positive.

Proportion of scores between the mean and z
> In that space between a given score and the mean, there will be a certain proportion of scores. This is the probability that a random score will fall in the area defined by the mean and any z.

Proportion of scores in the larger area

> You could also describe the area between any given z and a z of 4.00, or the end of the curve.

> The curve doesn't really ever end, theoretically, but a z score of 4.00 will come very close to including 100 percent of the scores.

> There are two ends of the curve, though. Unless your z is 0.0, the distance between the z and one end of the curve will be greater than the distance between the z and the other end. This column refers to the area between the z and that furthest end of the curve, and the value in this column is the proportion of scores that will fall in that space. In other words, it is the chance that a random person will produce a score in that area.

Proportion of scores in the smaller area

> This column refers to the area between the z and that closest end of the curve. It is the proportion of scores that will fall in that space.

Estimating the Chance of Scoring Above or Below Any Score

If you need to know your chances of getting into your college of choice, identify the necessary score you need to beat, also known as the *cut score*, on that school's admissions tests. Once you know the score, find out the mean and standard deviation for the test. (All of this info is probably on the Web.) Convert your raw score to a z score **[Hack #26]**, and then find that z score, or something close to it, in Table 3-2.

Determine whether the cut score is above the mean:

- If it is, look at the "Proportion of scores in the smaller area" column. That represents your chances of scoring at or above that cut score, and your chances of getting in.

- If the cut score is below the mean (unlikely, but for the sake of completely training you on how to use this tool), identify "Proportion of scores in the larger area." That's the proportion of students being accepted and, thus, your chances, all things being equal.

For the chances of scoring *below* a given score, the process is the opposite of the options just mentioned. The chance of getting below a specific cut score that is below the mean is shown in the "smaller area" column. The chance of scoring below a given cut score that is above the mean is shown in the "larger area" column.

Estimating the Chance of Scoring Between Any Two Scores

The chances of getting a score within any range of scoringscores can be determined by looking at the proportion of scores that will normally fall in that range.

If you want to know what proportion of scores falls between any two points under the curve, define those points by their z score and figure out the relevant proportion. Depending on whether both scores fall on the same side of the mean, one of two methods will give you the correct proportion between those points:

- If the z scores are on the same side of the curve, look up the proportion of scores in either the "larger area" or "smaller area" column for both z scores and subtract the lower value from the higher value.

- If the z scores fall on both sides of the mean with the mean between them, use the "Proportion of scores between the mean and z" column. Look up the value for both scores and add them together.

Producing Percentile Ranks

A third use of the table is to compute percentile ranks. You can read more about such *norm-referenced* scores in "Produce Percentiles" **[Hack #24]**. For scores above the mean, the percentile rank is "Proportion of scores between the mean and z" plus .50. For scores below the mean, the percentile rank is "Proportion of scores in the smaller area."

Determining Statistical Significance

Another use for these sorts of tables is to assign statistical significance **[Hack #4]** to differences in scores. By knowing the proportion of scores that will fall a certain distance from each other or further, you can assign a statistical probability to that outcome.

More usefully, other statistical values such as correlations and proportions can be converted to z scores, and this table can be used to compare those values to zero or to each other.

Why It Works

"See the Shape of Everything" **[Hack #23]** provides a good picture of the normal curve. However, just by looking at the way these values change in Table 3-2, you can get a good sense of the normal distribution's shape. Near the mean, where the rows have smaller z scores, a goodly proportion of scores will fall. As you move further and further away from the mean, it takes larger and larger areas of the curve to contain the same proportion of scores.

For example, it takes a jump from a z of 1.65 to 4 just to cover that last 5 percent of the distribution. Near the mean, though, it requires only a jump from $z = .12$ to $z = .25$ to cover 5 percent of scores. The table demonstrates how common it is to be common and how rare it is to be scarce.

See Also

- You will be able to compute your own exact areas under the normal curve by using this web site: *http://www.psychstat.missouristate.edu/introbook/sbk11m.htm*. A good discussion and some interactive calculators are part of this site maintained by David Stockburger. When you visit, don't be confused by words like *Mu* and *Sigma*. That's stats talk for mean and standard deviation, respectively.

HACK #26 Give Raw Scores a Makeover

A raw score on a test has little or no meaning. Change that pitiful raw score to a "z score," though, and you will scarcely believe how much information is crammed into that one little super number.

It is surprising how little information is conveyed by that single raw score plastered at the top of something like a high school test. Here's what I mean. If I come home from school and tell my mom that I got a 16 on the big exam in school today, she'll probably say a few things, including "Why are you still living at home at age 42?" and "That's nice, dear. Is 16 good?"

When you just tell someone a raw score, very little real information has been shared. You don't know if 16 *is* good. You don't know if 16 is relatively high or low. Did most people get a 16 or higher, or did most people get something less than 16? Even if we know the range of scores on that test and the points possible and so on, we still can't compare performance on that test to performance on the past test or the next test or a test on some other subject. Raw scores are virtually meaningless.

Don't fret! You can still understand your performance and the performances of others. You can still make selection decisions and compare performance across people and across tests. There is still hope!

Raw scores can be changed into a new number that does all the things that that 97-pound weakling, the raw score, could never do. Raw scores can be transformed into a super number: a *z score*. Unlike a raw score, a *z* tells you whether the performance is above or below average, and how far above or below average it is. A *z* also allows you to compare performance across tests and occasions, and even between people.

Calculating z Scores

A *z* score is a raw score that has been transformed in such a way that the new number indicates how far above or below the mean the raw score is.

Here's the equation:

$$z = \frac{\text{raw score} - \text{mean}}{\text{standard deviation}}$$

To change a raw score into a z, subtract the mean from it and then divide by the standard deviation. The standard deviation of a distribution is the average distance of each score from the mean [Hack #2].

Understanding Performance

z scores typically take on a range of values between −3 and +3. Examine the top part of the z score equation and you might notice the following:

- If the raw score is greater than the mean, the z will be positive.
- If the raw score is below the mean, the z will be negative.
- If the raw score is exactly the mean, the z will be 0.

> z scores tend to range between −3 and +3 because the *normal distribution* of scores is typically just six standard deviations wide [Hack #23].

Smart measurement professionals use the z score trick when they report results. Instead of supplying raw scores, all you see are scores based on z scores, known generically as standardized scores [Hack #27]. These standardized scores have known stable characteristics. Therefore, if you know these scores' characteristics (their mean and standard deviation), you can turn them back into z scores and know how you did compared to other people.

To see how to use this formula to reveal hidden information about your performance, let's use the example of ACT tests. The American College Test is taken by juniors in many high schools across the U.S. and is required by many colleges for admission. It is a test of achievement and ability believed to predict performance in college.

Scores on any portion of the test range from 1 to 36. Though the actual test's descriptive statistics have drifted over the last few decades (as performance has improved), the official ACT mean is often reported as 18 with a standard deviation of 6. Imagine three students take the ACT and receive three different scores. We could use the mean and standard deviation from the ACT score distribution to transform them to z scores, as shown in Table 3-3.

Table 3-3. *Transforming raw scores to z scores*

Student	ACT score	raw score − mean / standard deviation	Z score
Zack	14	$\dfrac{14 - 18}{6} = -\dfrac{4}{6}$	−.67
Taylor	18	$\dfrac{18 - 18}{6} = \dfrac{0}{6}$	0.00
Isaac	24	$\dfrac{24 - 18}{6} = \dfrac{6}{6}$	1.00

Zack's z is negative, so we know he scored below average. He scored about two-thirds of a standard deviation below the mean. Taylor's z of 0.00 means he performed average compared to others who have taken the ACT over the years. Isaac did the best, scoring a full standard deviation above the mean.

> The actual ACT mean and standard deviation changes every year the test is given. The *real* mean and standard deviation for the last few years has been around a mean of 21 and a standard deviation of about 4.5.

Identifying the Rarity of Your Performance

Though knowing how you scored in comparison with others who took the test is more useful than just knowing a raw score, the real interpretative power of z scores comes from its relationship to the normal curve. Figure 3-3 is a chart of the normal distribution, similar to the one shown in "See the Shape of Everything" [Hack #23].

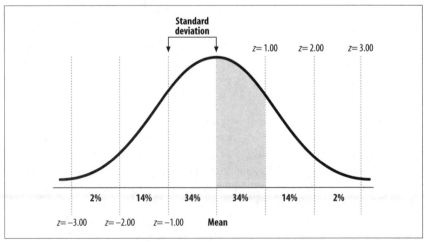

Figure 3-3. *z scores and the normal curve*

The difference between the figure in "See the Shape of Everything" [Hack #23] and this one is that instead of showing the distance of each standard deviation from the mean, Figure 3-3 shows those values as z scores. By using knowledge of areas under the normal curve, you can learn even more from a z score. If the scores are normally distributed, there is a great deal you can say about the probability of scores in a certain range occurring.

The scores for the students shown in Table 3-3 can also be interpreted as the number of students they did better (or worse) than. Taylor's z of 0.00 means he did better than 50 percent of students. The kids' scores can also be expressed in a probabilistic sense. There was a 50 percent chance that Taylor would get a z of 0.00 or better. There is only a 16 percent chance of getting a z of 1.00 or better on any test, so Isaac did well compared to other students who took the test.

Why It Works

If converting raw scores to z scores so we can compare people to each other makes some sense to you, then you are not alone. For the last 100 years in the world of educational measurement, social scientists (and anyone who must evaluate human performance) have been attracted to the simplicity of *norm-referenced* interpretations. If we aren't sure what the score on a test really means, we can at least compare your score to how everyone else has done. We at least know whether you have *more* or *less* of whatever it is we just measured than other people have.

The alternative way to interpret educational and psychological scores is *criterion-referenced*. That approach requires knowing more about the trait or content that we have just measured and deciding beforehand how much is enough. Criterion-referenced measurement allows for everyone to get the same score as long as they meet the same criteria. The former approach has been and continues to be the most popular interpretative method, while the latter has just recently started to catch on.

H A C K
#27
Standardize Scores

Surprisingly, none of those well-known high-stakes tests, such as the SAT or ACT or intelligence tests, ever reports your raw score. Instead, test reports have transformed that useless number into a more meaningful score, one that can be used to understand your performance compared to everyone else who ever took the same test. Once you understand "standardized" scores, you can calculate them yourself and even invent your own.

"Give Raw Scores a Makeover" [Hack #26] discusses the superpowers of z scores. These standardized scores take meaningless raw scores and add all sorts of information to them. That's all well and good, and anyone using this book can interpret z scores and make decisions based on that information.

If you want to interpret many score reports, though (such as those SAT results you just got), you will not see a z score reported anywhere, but instead some weirdo customized standardized score, used only by that company, which is kind of like a z score but different enough to be meaningless for the uninitiated.

Never fear. Here are the tools you need to both interpret these strange standardized scores and, if you want, even create your own (for when you report scores to other people from your own weirdo test that is just about to sweep the nation and make you as rich as Mr. ACT or Ms. IQ or whoever makes money from our test-based society).

Problems with z Scores

There is a certain, shall I say, *ugliness* to z scores that prevents their widespread use when reporting performance to test takers or their parents or the colleges and employers who are considering them. Instead, most test companies use the z score as the first step in creating a more attractive standardized score, which is then reported.

A raw score is transformed into a z score using this formula:

$$z = \frac{\text{raw score} - \text{mean}}{\text{standard deviation}}$$

As described in greater detail in "Give Raw Scores a Makeover" **[Hack #26]**, this equation creates z scores that tend to range between –3.00 and +3.00, with 0.00 as the average and a standard deviation equal to one. Though very useful as a tool for interpreting test performance, people don't like these numbers when they see them because of a few problems:

- It can be negative. In fact, half of all z scores will be negative. It is hard to convince people who take tests that a negative score can be anything but bad news.

- A score of 0.00 is the average score! If we can't explain to people that a negative number isn't necessarily a bad thing, imagine trying to convince parents that we expect little Billy to get zero on the big test and we are pleased when he does.

- The highest score you can expect is a 3.00, and only 1 out of a 100 test takers will ever get that. It seems like an awful lot of hard work in test preparation just to get a measly 3!

Measurement folks have searched for and found other standardized scales to report test performance that have more pleasing properties. The trick is to start with a z score, and then convert it onto some other scale with a mean and standard deviation that is friendlier.

Creating and Interpreting T Scores

One problem with z scores is that the mean is zero. Reporting zero as if it is an okay thing rubs some teachers, parents, and students the wrong way. We can solve that problem by moving down the alphabet form a z to a T.

T scores are a transformation of z scores into a new distribution that has a mean of 50 and a standard deviation of 10. The equation for a T score uses this backwards transformation approach. Here's the T score formula:

$$T = z(10) + 50$$

So, if little Billy's performance on a big test is average and he gets a z score of 0.00, instead of reporting that frightening score to his parents, we can transform it into a T:

$$T = 0.00(10) + 50 \qquad T = 0.00 + 50 \qquad T = 50$$

and report that Billy scored a 50. Congratulations! To make the score meaningful, a good teacher or school counselor would explain that T scores range from about 20 to 80, and 50 is average.

T scores are used on some test reports as a better alternative to z scores. Scores cannot be negative, and the mean is a more substantial-seeming 50.

One popular test that reports scores using the T score distribution is the Minnesota Multiphase Personality Inventory-II, a psychological test that measures depression, schizophrenia, and so on. Mean scores on each MMPI-II subscale are 50, with a standard deviation of 10. By putting each subtest score on the same scale, you can compare across traits and develop a profile of scores to understand the test taker more completely.

Creating Customized Standardized Scores

Test developers have found other ways of reporting standard scores. Table 3-4 lists many of the best-known high-stakes tests that most people have taken or will take someday.

Table 3-4. *Common standardized score distributions*

Test	Typical score range	Mean	Standard deviation
z scores	–3.00 to 3.00	0	1
T scores	20 to 80	50	10
American College Test (ACT)	1 to 36	18	6
SAT	200 to 800	500	100
Graduate Record Exam (GRE)	200 to 800	500	100
Graduate Management Admission Test (GMAT)	200 to 800	500	100
Law School Admission Test (LSAT)	120 to 180	150	10
Medical College Admission Test (MCAT)	1 to 15	8	2.5
Wechsler Intelligence Scales (IQ Test)	55 to 145	100	15
Stanford-Binet Intelligence Test (IQ Test)	52 to 148	100	16

Because test performance is normally distributed, you can interpret any of these scores by placing it against the normal curve and seeing whether your performance was average, unusually low, or unusually high [Hack #23].

Create Your Own Standardized Score

For fun, you can create your own standardized score distribution with any mean and standard deviation you wish. Don't like your SAT score of 350? Transform it into a score within a distribution of your choosing.

Imagine, for example, that you'd prefer a distribution with a mean of 752,365 and a standard deviation of 216,456 (and who wouldn't?). Let's call this distribution the *Frey Score Distribution*. Generalizing the *T* score formula, you could transform your SAT score of 350 into a *Frey* score. Remember, you have to start with the z score for an SAT score of 350:

$$z = \frac{\text{raw score} - \text{mean}}{\text{standard deviation}} = \frac{350 - 500}{100} = \frac{-150}{100} = -1.50$$

and then transform it into a *Frey* score:

$$\text{Frey} = -1.50(216,456) + 752,365 = -324,684 + 752,365 = 427,681$$

Now, doesn't a score of 427,681 sound better than a score of 350? Because you know the mean of the *Frey* distribution, the interpretation of both scores is the same; they are still below average, and they are still 1½ standard deviations below the mean. You haven't changed reality, just the numbers you use to describe it.

Why It Works

The distribution of z scores has a mean of 0 and a standard deviation of 1. This is because of the equation used. By dividing a group of values by its standard deviation, the standard deviation of the new distribution is 1. By subtracting the mean from each score in a distribution, the new values distribute themselves around a mean of 0.

If we want the scores we use to have a particular mean and standard deviation of our own choosing, we can take each z score and reverse engineer it, replacing the mean of 0 with anything we want and the standard deviation of 1 with anything we want.

Understanding Norm-Referenced Scoring

We have talked about the information inherent in norm-referenced scoring and its intuitive appeal from a statistical perspective, but it is not the only way to produce meaningful scores, and it's not always the best method.

As discussed in "Give Raw Scores a Makeover" [Hack #26], there are really two philosophies from which you can choose when designing scoring systems and building tests:

Norm-referenced scoring
> Driven by the philosophy that to best understand performance on a task (such as acting in a movie or taking the ACT), the level of performance for one person should be compared to how other people performed

Criterion-referenced scoring
> Evaluates performance based on a set of criteria, such as a base of knowledge, a set of skills, instructional objectives, and diagnostic characteristics

If the norm-referenced approach makes sense to you, then you will want to use the tools presented here to interpret your performance on these common standardized tests.

Ask the Right Questions

HACK #28

If you are a classroom teacher, a job interviewer, or in any situation where you want to measure someone's understanding, you have a variety of ways to ask a question. Here are some tools from the science of measurement that allow you to ask the right question in the right way.

For more than a hundred years, classrooms have been an environment of questions and answers. Outside of school, tests are more and more common in the workplace and in hiring decisions. Even in my free time, I can't pick up a *Cosmo* without having to respond to a relationship quiz about whether I am "friendly" or "frosty" when it comes to meeting people at parties. (I'm frosty. Want to make something of it?)

Many professions have to ask good questions or write good tests:

- Teachers ask students questions while lecturing or one-on-one in private conferences to assess student understanding.
- Trainers write questions to evaluate the effectiveness of workshops.
- Personnel officers develop standard questions to measure applicants' skills.

Anyone who ever has to assess how much someone else knows is faced with the dilemma of deciding what sort of question to ask to really get to the heart of the matter. This hack provides solutions to the two most common problems when writing tests or designing questions meant to measure knowledge or understanding:

- How do I construct a good question?
- What should I ask about?

Constructing a Good Question

For measuring knowledge quickly and efficiently, it is hard to beat the *multiple-choice* item as a question format.

> Multiple-choice questions are a type of item that presents respondents with a question or instruction (called the *stem*), and then asks them to select the correct answer or response from a list of answer options. These types of items are sometimes referred to as *selection* items because people select the answer.

To give us the right terms to use as we talk about how to write a good multiple-choice item, a quick primer is in order.

Here is an example of a multiple-choice item:

Who wrote *The Great Gatsby*?	←Stem
A. Faulkner	←Distractor
B. Fitzgerald	←Correct answer ("keyed" answer)
C. Hemingway	←Distractor
D. Steinbeck	←Distractor

As you see, each part of the question has a name. The correct answer is called the *correct answer* (how's that for scientific jargon?), and wrong answers are called *distractors*.

Not much, but some real-world research has been done on the characteristics of multiple-choice items and how to write good ones. To write good multiple-choice items, follow the following critical item-writing guidelines from this research:

Include 3 to 5 answer options
> Items should have enough answer options that pure guessing is difficult, but not so many that the distractors are not plausible or the item takes too long to complete.

Do not include "All of the Above" as an answer option
> Some people will guess this answer option frequently, as part of a test-taking strategy. Others will avoid it as part of a test-taking strategy. Either way, it does not operate fairly as a distractor. Additionally, to evaluate the possibility that "All of the Above" is correct requires analytical abilities that vary across respondents. Measuring this particular analytic ability is likely not the targeted goal of the test.

Do not include "None of the Above" as an answer option
> This guideline exists for the same reasons as the previous guideline. Additionally, for some reason, teachers do tend to create items where "None of the Above" is most likely to be the correct answer, and some students know this.

Make all answer options plausible
> If an answer option is clearly not correct because it does not seem related to the other answer options, it is from a content area not covered by the test, or the teacher is obviously including it for humorous reasons, it does not operate as a distractor. Students are not considering the distractor, so a four-answer-option question is really a three-answer-option question and guessing becomes easier.

Order answer options logically or randomly
> Some teachers develop a tendency to write items where a certain answer option (e.g., B or C) is correct. Students might pick up on this with a given teacher. Additionally, some courses on doing well on standardized multiple-choice tests suggest this technique as part of a test-taking strategy. Teachers can control for any tendencies of their own by placing the answer options in an order based on some rule (e.g., shortest to longest, alphabetical, chronological).

> Another solution to this ordering problem is for teachers to scroll through the first draft of the test on their word processors and attempt to randomize the order of answer options. Computerized randomization is the solution, of course, for commercial standardized test developers as well.

Make the stem longer than answer options

An item is processed more quickly if the bulk of the reading is in the stem, followed by brief answer options.

 Because longer stems followed by shorter answer options allows for easier processing for test takers, a good multiple-choice item should look like this:

==

====================

====================

====================

====================

Do not use negative wording

Some students read more carefully or process words more accurately than others, and the word "not" can easily be missed. Even if the word is emphasized so no one can miss it, educational content tends not to be learned as a collection of non-facts or false statements, but is likely stored as a collection of positively worded truths.

Make answer options grammatically consistent with stem

For example, if the grammar used in the stem makes it clear that the right answer is a female or is plural, make sure that all answer options are female or plural.

Use complete sentences for stems

If a stem is a complete question ending with a question mark, or a complete instruction ending with a period, students can begin to identify the answer before examining answer options. Students must work harder if stems end with a blank or a colon, or if it's simply an uncompleted sentence. More processing increases chances of errors.

Asking a Question at the Right Level

Identifying the right level of question to ask is the second major problem that must be overcome when creating tests. Some questions are easy; they only assess one's ability to recall information and indicate a fairly low level of knowledge. Other questions are more difficult and require a response that combines existing knowledge or applies it to a new problem or situation. Because different levels of questions measure different levels of understanding, the right question must be asked at the right level for anything useful to be gained from the enterprise.

A smart fellow and educational researcher, Benjamin Bloom, writing in the 1950s, suggested a way of thinking about questions and the level of understanding required to respond correctly. His classification system has become known as *Bloom's Taxonomy*, a classification system of educational objectives based on the level of understanding necessary for achievement or mastery. Bloom and colleagues have suggested six different cognitive stages in learning. They are, in order from lowest to highest:

1. *Knowledge*
   Ability to recall words, facts, and concepts

2. *Comprehension*
   Ability to understand and communicate about a topic

3. *Application*
   Ability to use generalized knowledge to solve an unfamiliar problem

4. *Analysis*
   Ability to break an idea into parts and understand their relationship

5. *Synthesis*
   Ability to create a new pattern or idea out of existing knowledge

6. *Evaluation*
   Ability to make informed judgments about the value of new ideas

Choosing the right cognitive level. Let's use teachers as an example of how to think about what level of questions you want. Teachers choose the appropriate cognitive level for classroom objectives, and a quality assessment is designed to measure how well those objectives have been met. Most items written by teachers, and those on prewritten tests packaged with textbooks and teaching kits, are at the *knowledge* level. Most researchers consider this unfortunate, because classroom objectives should be (and usually are) at higher cognitive levels than simply memorizing information.

When new material is being introduced, however (at any age—preschool through advanced professional training), an assessment probably should include at least a check that basic new facts have been learned. When teachers decide to measure beyond the knowledge level, the appropriate level for items depends on the developmental level of students. The cognitive level of students, particularly their ability to think and understand abstractly, and their ability to solve problems using multiple steps, should determine the best level for classroom objectives, and, therefore, the best level for test items. Researchers believe that teachers should test over what they teach, in the same way that they teach it.

So, any time you find yourself wanting to assess the knowledge hidden inside someone's head, think about what level of understanding you want to assess. Is basic memorized knowledge enough? If so, then the *knowledge* level is the appropriate level for a question. Do you want to know whether your job applicant can use her knowledge to solve problems she has never experienced before? Ask a question at the *application* level, and she will have to demonstrate that ability.

Designing questions at different cognitive levels. Follow the guidelines in Table 3-5 for creating items or tasks at each level of Bloom's Taxonomy.

Table 3-5. Questions at different cognitive levels

Bloom's level	Question characteristics	Example question or task
Knowledge	Requires only rote memory ability and such skills as recall, recognition, and repeating back	Who wrote *The Great Gatsby*? A. Faulkner B. Fitzgerald C. Hemingway D. Steinbeck
Comprehension	Requires skills such as paraphrasing, summarizing, and explaining	What is a prehensile tail?
Application	Requires skills such as performing operations and solving problems, and includes words such as *use*, *compute*, and *produce*	If a farmer owns 40 acres of land and buys 16 acres more, how many acres of land does she own?
Analysis	Requires skills such as outlining, listening, logic, and observation, and uses words such as *identify* and *break down*	Draw a map of your neighborhood and identify each home.
Synthesis	Requires skills such as organization and design, and includes words such as *compare* and *contrast*	Based on your understanding of the characters, describe what might happen in a sequel to *Flowers for Algernon*.
Evaluation	Requires skills such as criticism and forming opinions, and includes words such as *support* and *explain*	Which musical film performer was probably the best athlete? Defend your answer.

When to use Bloom's Taxonomy. There is an implied hierarchy to Bloom's categories, with *knowledge* representing the simplest level of cognition and *evaluation* representing the highest and most complex level. Anyone writing questions to assess knowledge can write items for any given level. Teachers can identify the level of chosen classroom objectives and create assessments to match those levels. With objectively scored item formats, it is fairly simple to tap lower levels of Bloom's taxonomy and more difficult, but not impossible, to measure at higher levels.

You should not worry too much about the fine distinctions between the six levels as defined by Bloom. For example, *comprehension* and *application* are commonly treated as synonymous, as it is the ability to apply what is learned that indicates comprehension. Most testing theorists and classroom teachers today pay the most attention to the distinction between the *knowledge* level and all the rest of the levels. Most teachers, except at introductory stages of brand new areas, prefer to teach and measure to objectives that are above the *knowledge* level.

See Also

- Here's something a little more scholarly that I wrote with some colleagues: Frey, B.B., Petersen, S.E., Edwards, L.M., Pedrotti, J.T., and Peyton, V. (2005). "Item-writing rules: Collective wisdom." *Teaching and Teacher Education, 21*, 357–364.

- For a good review of item-writing rules, check out Haladyna, T.M., Downing, S.M., and Rodriguez, M.C. (2002). "A review of multiple-choice item-writing guidelines for classroom assessment." *Applied Measurement in Education, 15*(3), 309–334.

- The influential ideas in Bloom's taxonomy were introducd in Bloom, B. S. (Ed.). (1956). *Taxonomy of educational objectives: The classification of educational goals. Handbook 1. Cognitive domain.* New York: McKay.

- Bloom, B.S., Hastings, J.T., and Madaus, G.F. (1971). *Handbook on formative and summative evaluation of student learning.* New York: McGraw-Hill.

- Phye, G.D. (1997). *Handbook of classroom assessment: Learning, adjustment, and achievement.* San Diego, CA: Academic Press.

Test Fairly

HACK #29

Classroom teachers frequently create their own tests to measure their students' learning. They often worry whether their tests are too hard or too easy and whether they measure what they are supposed to measure. Item analysis tools provide the solutions to teachers' concerns.

Classroom assessment is perhaps the single most common activity in the modern schoolroom. Teachers are always making and grading tests, students are always studying for and taking tests, and the whole process is meant to support student learning. Tests must not be too hard (or too easy), and they must measure what the teacher wants them to measure. Test scores and grades are the way that teachers communicate with parents, students, and administrators, so the score at the top of the test needs to be fair. It must accurately reflect student learning, and it should be the result of a quality assessment.

Concerned teachers constantly work to improve their tests, but they are often working in the dark without solid data to guide them. What can a smart, caring teacher do to improve his tests or improve the validity of his grading? A family of statistical methods called *item analysis* can provide direction to teachers as they seek to develop fair assessments and grading.

Item Analysis

Item analysis is the process of examining classroom performance on individual test items. A classroom teacher might want to examine performance on parts of a test she has written, to see what areas are being mastered by her students and what areas need more review. A commercial test developer producing exams for nursing certification might want to know which items on his test are the most valid and which seem to measure something else and should therefore be removed.

In both cases, the developer of the test is interested in item difficulty and item validity. Though one example involves a high school teacher making tests for her own students, and the other example involves a large for-profit corporation, both developers are interested in the same types of data, and both can apply the same tools of item analysis.

Three Types of Classroom Assessment Problems

If you are a classroom teacher worried about your own assessments, there are three different types of questions that you probably need to answer. Fortunately, there are three item-analysis tools that will provide you with the three different types of information you need.

Are my test questions too hard? The difficulty of any specific test question can be calculated fairly easily using the formula for the *difficulty index*. You can produce a difficulty index for a test item by calculating the proportion of students taking the test that got that item correct. The larger the proportion, the more test takers who know the information measured by the item.

The term *difficulty index* is counterintuitive, because it actually provides a measure of how *easy* the item is, not the *difficulty* of the item. An item with a high difficulty index is an easy item, not a tough one.

How hard is too hard? You get to decide that yourself. Some teachers treat difficulty indices at .50 or below as too hard because most people missed the item. You might have higher standards. If you believe that most students should have learned the material and your difficulty index for an item suggests that a substantial portion of your class missed it, it might be too hard.

Is each test question measuring what it is supposed to? Measurement experts say that if a test item measures what it is supposed to, then it is valid [Hack #32]. The *discrimination index* is a basic measure of the validity of an item, in addition to its reliability. It measures an item's ability to discriminate between those who scored high on the total test and those who scored low.

Though there are several steps in its calculation, once computed, this index can be interpreted as an indication of the extent to which overall knowledge of the content area or mastery of the skills is related to the response on an item.

> A *discrimination* index is not so named because it suggests test *bias*. *Discrimination* is the ability to identify whether one who got an item correct is in a high-scoring group or a low-scoring group.

Why did my students miss a question? In addition to examining the performance of an entire test item, teachers are often interested in examining the performance of individual distractors (incorrect answer options) on multiple-choice items through *analysis of answer options*. By calculating the proportion of students who choose each answer option, teachers can see what sorts of errors students are making. Have they mislearned certain concepts? Do they have common confusions about the material?

To improve how well the item works from a measurement perspective, teachers also can identify which distractors are "working" and appear attractive to students who do not know the correct answer, and which distractors are simply taking up space and are not being chosen by many students.

To eliminate educated guesses that result in correct answers purely by chance, teachers and test developers want as many plausible distractors as is feasible. Analyses of response options allow teachers to fine-tune and improve items they might want to use again with future classes.

Conducting Item Analyses and Interpreting Results

Here are the procedures for the calculations involved in item analysis, using data for an example item. For this example, imagine a classroom of 25 students who took a test that included the item in Table 3-6 (keep in mind, though, that even large-scale standardized test developers use the same procedures for tests taken by hundreds of thousands of people).

The asterisk for the answer options in Table 3-6 indicates that B is the correct answer.

Table 3-6. Sample item for item analysis

Answer to question: "Who wrote *The Great Gatsby?*"	Number of students who chose each answer
A. Faulkner	4
B. Fitzgerald*	16
C. Hemingway	5
D. Steinbeck	0

To calculate the difficulty index:

1. Count the number of people who got the correct answer.
2. Divide by the total number of people who took the test.

On the item shown in Table 3-6, 16 out of 25 people got the item right:

16 / 25 = .64

Difficulty indices range from .00 to 1.0. In our example, the item had a difficulty index of .64. This means that 64 percent of students knew the answer.

If a teacher believes that .64 is too low, there are a couple of actions she can take. She could decide to change the way she teaches to better meet the objective represented by the item. Another interpretation might be that the item was too difficult or confusing or invalid, in which case the teacher can replace or modify the item, perhaps using information from the item's discrimination index or analysis of response options.

To calculate the discrimination index:

1. Sort your tests by total score, and create two groupings of tests: the *high scores*, made up of the top half of tests, and the *low scores*, made up of the bottom half of tests.
2. For each group, calculate a difficulty index for the item.
3. Subtract the difficulty index for the low scores group from the difficulty index for the high scores group.

Imagine that in our example 10 out of 13 students (or tests) in the high group and 6 out of 12 students in the low group got the item correct. The high group difficulty index is .77 (10/13) and the low group difficulty index is .50 (6/12), so we can calculate the discrimination index like so:

.77 − .50 = .27

The discrimination index for the item is .27. Discrimination indices range from −1.0 to 1.0. The greater the positive value (the closer it is to 1.0), the stronger the relationship is between overall test performance and performance on that item.

If the discrimination index is negative, that means that, for some reason, students who scored low on the test were more likely to get the answer correct. This is a strange situation, and it suggests poor validity for an item or that the answer key was incorrect. Teachers usually want each item on the test to tap into the same knowledge or skill as the rest of the test.

> The formula for the discrimination index is such that if more students in the high-scoring group chose the correct answer than did students in the low-scoring group, the number is positive. At a minimum, then, a teacher would hope for a positive value, because that would indicate that knowledge resulted in the correct answer.

We can use the information provided in Table 3-6 to look at the popularity of different answer options, as shown in Table 3-7.

Table 3-7. Item analysis of "Who wrote The Great Gatsby?"

Answer	Popularity of options	Difficulty index
A. Faulkner	4/25	.16
B. Fitzgerald*	16/25	.64
C. Hemingway	5/25	.20
D. Steinbeck	0/25	.00

The analysis of response options shows that students who missed the item were about equally likely to choose answer A and answer C. No students chose answer D, so answer option D does not act as a distractor. Students are not choosing between four answer options on this item; they are really choosing between only three options, since they are not even considering answer D.

This makes guessing correctly more likely, which hurts the validity of an item. A teacher might interpret this data as evidence that most students make the connection between *The Great Gatsby* and Fitzgerald, and that the students who don't make this connection can't differentiate between Faulkner and Hemingway very well.

Suggestions for Item Analysis and Test Fairness

To improve the quality of tests, item analysis can identify items that are too difficult (or too easy, if a teacher has that concern), don't differentiate between those who have learned the content and those who have not, or have distractors that are not plausible.

If you as a teacher have concerns about test fairness, you can change the way you teach, change the way you test, or change the way you grade the tests:

Change the way you teach
> If some items are too hard, you can adjust the way you teach. Emphasize unlearned material or use a different instructional strategy. You might specifically modify instruction to correct a confusing misunderstanding about the content.

Change the way you test
> If items have low or negative discrimination values, they can be removed from the current test, and you can remove them from the pool of items for future tests. You can also examine the item, try to identify what was tricky about it, and change the item. When distracters are identified as being nonfunctional (no one picks them), teachers can tinker with the item and create a new distracter. One goal for a valid and reliable test is to decrease the chance that random guessing could result in credit for a correct answer. The greater the number of plausible distracters, the more accurate, valid, and reliable the test typically becomes.

Change the way you grade
> You might use item analysis information to decide that the material was not taught and, for the sake of fairness, remove the item from the current test and recalculate scores. The simplest way for real classroom teachers to do this is to simply count the number of *bad* items on a test and add that number to everyone's score. This is not technically the same as rescoring the test as if the item never existed, but this way students still get credit if they got a hard or tricky item correct, which seems fairer to most teachers.

These concerns that teachers have about the quality of their tests are not much different than the research questions that scientists ask. Just like scientists, teachers can collect data in their classroom, analyze the data, and interpret results. They can then decide, based on their own personal philosophies, how to act on those results.

HACK #30 Improve Your Test Score While Watching Paint Dry

If you don't like the score you just got on that important high-stakes test, maybe you should take the test again. Or should you?

We've already discussed how to measure anything precisely by applying concepts of reliability [Hack #6]. *Reliability* is the consistency with which a test assesses some outcome. In other words, a reliable test produces a stable score, and an unreliable test does not. Because tests that are less than perfectly reliable produce scores at least partly due to random chance, their scores can move around in ways that statisticians can predict. Because your test score when you retake a test will tend to move toward the average score on that test, this effect is called *regression toward the mean*.

When you take a high-stakes test such as the SAT, ACT, GRE, LSAT, or MCAT, you often have the option of retaking it to try to improve your score. Your decision on whether it is worth the time, hard work, and money to try to improve your test score should be made with an understanding of the test's reliability and how much change is possible simply through regression to the mean.

Regressing to the Mean

First, let's make regression to the mean occur, so you'll believe that scores can change in a predictable direction for no reason other than the characteristics of the normal curve [Hack #23]. Seeing is believing, and I hope to make this invisible magical phenomenon happen before your eyes.

Give the true/false quiz shown in Table 3-8 to 100 of your closest friends. Well, OK, maybe 10 people, counting you. 1,000 would be even better, but I just need enough to prove to you that this regression thing happens. As we proceed, keep in mind that if we had 100 or 1,000 takers of this very difficult (or very easy) test, the results would be even more convincing.

Oh, and for this test, you don't have to see the actual questions themselves. Scores will change on this test without any change in the construct that is being measured [Hack #32]. So, all you can do on this quiz is guess. Because they are true/false questions, you will have a 50 percent chance of getting any question correct, and the average performance for your group of 10 test takers (or 100 if you are really serious about this...can you do at least 30 maybe?...anyone?) should be a score of 5 out of 10.

Table 3-8. Advanced Quantum Physics Quiz

Question	Circle Your Answer
1.	True or False
2.	True or False
3	True or False
4.	True or False
5.	True or False
6.	True or False
7.	True or False
8.	True or False

Table 3-8. Advanced Quantum Physics Quiz (continued)

Question	Circle Your Answer
9.	True or False
10.	True or False

Administer the *Advanced Quantum Physics Quiz* to all the people you were able to get. And when you and the others take this quiz, don't cheat by looking at the answer key, even though it is only inches away from your eyes right now (in Table 3-9)!

Table 3-9. Answer key for the Advanced Quantum Physics Quiz

1. True	2. True	3. False	4. False	5. True
6. False	7. False	8. True	9. True	10. False

Collect the completed tests (make sure they put their names on them) and score them up, using the answer key in Table 3-9.

Now, pick your highest scorer (this represents someone like you, perhaps, who scores higher than average on standardized tests such as the SAT) and the lowest scorer (this represents someone not like you, perhaps, who scores lower than average). Give these two people the quiz again (without them seeing the correct answers) and score them again.

Here's where *regression to the mean* kicks in. I am pretty sure—without knowing you or your friends or what their answers are—of two things:

- The person who scored lowest the first time will score higher than he did before.
- The person who scored highest the first time will score lower than she did before.

If it worked, then aha! I told you so. If it didn't work, I told you I was only "pretty sure." With a larger sample, it is much more likely to work.

Why It Works

What we expect to happen with the two scores is that all the test scores that are below 5 (or whatever your test mean was) would move up toward the mean, and those scores above 5 would move down toward the mean. This may or may not have happened with your two scores, but it is the most probable outcome.

Remember this was a test in which knowledge had no effect on scores. Scores were due entirely to chance both times. This effect occurs with real tests, though, even when knowledge does influence your score. That's because no real test is perfectly reliable, and chance plays some role in performance on every test. This demonstration just exaggerated the effect by presenting a test in which chance accounts for 100 percent of the test taker's score.

So, why are scores likely to change and move closer to the mean on second occasions? In the long run, with 100 or 1,000 sets of test scores, we would expect the outcomes to be something like the normal distribution. Just like flipping a coin (which can come up heads or tails, with a 50 percent chance of either), probabilities are associated with particular outcomes on a true/false test (or any test, for that matter). Table 3-10 shows the possible scores and the likelihood of a test taker receiving them for the *Advanced Quantum Physics Quiz*.

Table 3-10. Likely quiz score distribution

Score	Probability
0	0.001
1	0.010
2	0.044
3	0.117
4	0.205
5	0.246
6	0.205
7	0.117
8	0.044
9	0.010
10	0.001

Why would more extreme scores become less extreme with repeated testing? Look at the likelihood of getting two extreme scores (such as a score of 2 and then another score of 2) versus getting a score of 2 (probability = .044), and then a score of 4 (probability = .205). It's almost five times as likely that a person with a 2 the first time will score a 4 on a second administration. It is almost 95 percent certain that he will score higher than 2 (1 − .044 − .010 − .001 = .945).

> The phrase "regression toward the mean" gets its name from the famous (and half cousin to Charles Darwin) Francis Galton, who studied the heights of parents and their children. He found that the average height of the children was closer to the mean height of all children than to the mean of the average height of the children's parents. While Galton called this observation "regression toward mediocrity" (Galton was not known to be a diplomat), we're a bit kinder. It has nothing to do with genetics and everything to do with—you guessed it—statistics.

With this test, in which scores were entirely due to chance, there is a 65.6 percent chance of scoring at or very near the mean (combining probabilities of scores 4, 5, and 6). With most tests, which have a greater number of items and produce normal distributions, you have a 68 percent chance of scoring at or near the mean [Hack #23].

Predicting the Likelihood of a Higher Score

This is all very interesting, but how will it help you decide whether it is worth it to take a test a second time? Back to our original dilemma. Taking these important tests (such as college admissions tests) a second time takes more money, time, stress, and, perhaps, preparation, so one needs to be strategic in deciding when to try again.

 Of course, you can do better on a test by actually increasing your level of whatever knowledge the test is measuring. You are likely to score higher if you prepare for an exam through study, taking practice exams or preparation courses, and so on. If you score very low, though, you are likely to do better without having done anything between test administrations, just because of regression to the mean. You can watch paint dry between testing times and your score will still probably increase. Lucky dog!

The likelihood that you will do better on a test by just taking it a second time depends on two things: your score the first time and the reliability of the test.

Your score

Because scores are likely (by chance alone) to move toward the mean, the chance of you doing better given a second chance depends on whether your first score is below or above the mean. Think of the mean as that big sucking sound you hear, pulling all the scores along a distribution towards it. Scores below the mean are more likely to increase than are scores above the mean.

Test reliability

Measurement statisticians use a number for reliability, which represents the proportion of score variability that is *not* due to chance. The higher the reliability, then, the less of a role chance will play in determining your score. Reliable scores are stable scores, and the super-sucking powers of the mean are no match for a reliable score.

Statisticians have developed a formula that you can apply to give you a good idea of how much wiggle room you have around your score. If there is plenty of room to grow, you might consider a second shot at it. A useful tool to use here is the *standard error of measurement*. Here's the formula for the standard error of measurement [Hack #6]:

$$\text{Standard Error} = \text{Standard Deviation}\sqrt{1 - \text{Reliability}}$$

Most standardized tests publish their levels of reliability and the expected standard deviation for the many hundreds of thousands of scores produced by the test during each administration. By plugging values for these tests into the standard error of measurement equation, one can get a general sense of the variation of scores from test to retest that might be possible without any real change in the person being measured.

However, even the standard error is misleading for extreme scores. Very low scores and very high scores are likely to move a greater distance by chance alone than the standard error would suggest. The further you are from normal, the harder it is to resist the gravitational forces of normal. Extreme scores cannot resist that pull, unless they are perfectly reliable.

In sum, here's some sound advice on how to decide whether to retake a test:

- If you scored very high, relatively speaking, but not as high as you would like, it is probably not worth the trouble to take the test a second time.

- If you scored very low (far below average), it is almost certain that you will score higher the second time. Try again. You might study a little this time, too.

—Neil Salkind

Establish Reliability

People who use, make, and take high-stakes tests have a vested interest in establishing the precision of a test score. Fortunately, the field of educational and psychological measurement offers several methods for both verifying that a test score is consistent and precise and indicating just how trustworthy it is.

Anyone who uses tests to make high-stakes decisions needs to be confident that the scores that are produced are precise and that they're not influenced much by random forces, such as whether the job applicant had breakfast that morning or the student was overly anxious during the test. Test designers need to establish reliability to convince their customers that they can rely on the results produced.

Most importantly, perhaps, when you take a test that will affect your admission to a school or determine whether you get that promotion to head beverage chef, you need to know that the score reflects your typical level of performance. This hack presents several procedures for measuring the reliability of measures.

Why Reliability Matters

Some basics, first, about test reliability and why you should seek out reliability evidence for important tests you take. Tests and other measurement instruments are expected to behave consistently, both *internally* (measuring the same construct behaving in similar ways) and *externally* (providing similar results if they are administered again and again over time). These are issues of *reliability*.

Reliability is measured statistically, and a specific number can be calculated to represent a test's level of consistency. Most indices of reliability are based on correlations [Hack #11] between responses to items within a test or between two sets of scores on a test given or scored twice.

Four commonly reported types of reliability are used to establish whether a test produces scores that do not include much random variance:

Internal reliability
Is performance for each test taker consistent across different items within a single test?

Test-retest reliability
Is performance for each test taker consistent across two administrations of the same test?

Inter-rater reliability
Is performance for each test taker consistent if two different people score the test?

Parallel forms reliability
Is performance for each test taker consistent across different forms of the same test?

Calculating Reliability

If you have produced a test you want to use—whether you are a teacher, a personnel officer, or a therapist—you will want to verify that you are measuring reliably. The methods you use to compute your level of precision depend on the reliability type you are interested in.

Internal reliability. The most commonly reported measure of reliability is a measure of internal consistency referred to as coefficient (or Cronbach's) alpha. *Coefficient alpha* is a number that almost always ranges from .00 to 1.00. The higher the number, the more internally consistent a test's items behave.

If you took a test and split it in half—the odd items in one half and the even items in the other, for example—you could calculate the correlation between the two halves. The formula for split-half correlations is the correlation coefficient formula **[Hack #11]** and is a traditional method for estimating reliability, though it is considered a bit old-fashioned these days.

Mathematically, the formula for coefficient alpha produces an average of correlations between all possible halves of a test and has come to replace a split-half correlation as the preferred estimate of internal reliability. Computers are typically used to calculate this value because of the complexity of the equation:

$$\text{alpha} = \frac{n}{n-1}\left(\frac{\text{SD}^2 - \sum \text{SD}_i^2}{\text{SD}^2}\right)$$

where n = the number of items on the test, SD = standard deviation of the test, Σ means to sum up, and SD_i = standard deviation of each item.

Test-retest reliability. Internal consistency is usually considered appropriate evidence for the reliability of a test, but in some cases, it is also necessary to demonstrate consistency over time.

If whatever is being measured is something that should not change over time, or if it should change very slowly, then responses from the same group should be pretty much the same if they were administered the same test on two different occasions. A correlation between these two sets of scores would reflect a test's consistency over time.

Inter-rater reliability. We can also calculate reliability when more than one person scores a test or makes an observation. When different raters are used to produce a score, it is appropriate to demonstrate consistency between them. Even if only one scorer is used (as with a teacher in a classroom), if the scoring is subjective at all, as with most essay questions and performance assessments, this type of reliability has great theoretical importance.

To demonstrate that an individual's score represents typical performance in these cases, it must be shown that it makes no difference which judge, scorer, or rater was used. The level for inter-rater reliability is usually established with correlations between raters' scores for a series of people or with a percentage that indicates how often they agreed.

Parallel forms reliability. Finally, we can demonstrate reliability by arguing that it doesn't matter which form of a test a person takes; she will score about the same. Demonstrating parallel forms reliability is necessary only when the test is constructed from a larger pool of items.

For example, with most standardized college admission tests, such as the SAT and the ACT, different test takers are given different versions of the test, made up of different questions covering the same subjects. The companies behind these tests have developed many hundreds of questions and produce different versions of the same test by using different samples of these questions. This way, when you take the test in Maine on a Saturday morning, you can't call your cousin in California and tell him specific questions to prepare for before he takes the test next week, because your cousin will likely have a different set of questions on his test.

When companies produce different forms of the same test, they must demonstrate that the tests are equally difficult and have other similar statistical properties. Most importantly, they must show that you would score the same on your Maine version as you would if you took the California version.

Interpreting Reliability Evidence

There are a variety of approaches to establishing test reliability, and tests for different purposes should have different types of reliability evidence associated with them. You can rely on the size of the reliability coefficients to decide whether a test you have made needs to be improved. If you are only taking the test or relying on the information it provides, you can use the reliability value to decide whether you trust the test results.

Internal reliability
> A test designed to be used alone to make an important decision should have extremely high internal reliability, so the score one receives should be very precise. A coefficient alpha of .70 or higher is most often considered necessary for a claim that a test is internally reliable, though this is just a rule of thumb. You decide what is acceptable for the tests you make or take.

Test-retest reliability
> A test used to measure change over time, as in various social science research designs, should display good test-retest reliability, which means any changes between tests are not due to random fluctuations in scores. An appropriate size for a correlation of stability depends on how theoretically stable a construct should be over time. Depending on its characteristics, then, a test should produce scores over time that correlate in the range of .60 to 1.00.

Inter-rater reliability

Inter-rater reliability is interesting only if the scoring is subjective, such as with an essay test. Objective, computer-scored multiple-choice tests should produce perfect inter-rater reliability, so that sort of evidence is typically not produced for objective tests. If an inter-rater correlation is used as the estimate of inter-rater reliability, .80 is a good rule of thumb for minimum reliability.

Sometimes, reliability across raters is estimated by reporting the percentage of time the two scorers agreed. With a *percentage agreement* reliability estimate, 85 percent is typically considered good enough.

Parallel forms reliability

Only tests with different forms can be described as having parallel forms reliability. Your college professor probably doesn't need to establish parallel forms reliability when there is only one version of the final, but large-scale test companies probably do.

Parallel forms reliability should be very high, so people can treat scores on any form of the test as equally meaningful. Typically, correlations between two forms of a test should be higher than .90. Test companies conduct studies in which one group of people takes both forms of a test in order to determine this reliability coefficient.

Before you take a high-stakes test that could determine which roads are open to you, make sure that the test has accepted levels of reliability. The type of reliability you'd like to see evidence of depends on the purpose of the test.

Improving Test Reliability

The easiest way to ensure a high coefficient alpha or any other reliability coefficient is to increase the length of your test. The more items asking about the same concept and the more opportunities respondents have to clarify their attitudes or display knowledge, the more reliable a total score on that test would be. This makes sense theoretically, but also increases reliability mathematically because of the formula used to calculate reliability.

Look back at the equation for coefficient alpha. As the length of a test increases, the variability for the total test score increases at a greater rate than the total variability across items. In the formula, this means that the value in the parentheses gets larger as a test gets longer. The n/n-1 portion also increases as the number of items increases. Consequently, longer tests tend to produce higher reliability estimates.

Why It Works

Correlations compare two sets of scores matched up so that each pair of scores describes one individual. If most people perform consistently—each of their two scores is high, low, or about average when compared to other individuals, or a high score on one test matches consistently with a low score on another—the correlation will be close to 1.00 or –1.00.

An inconsistent relationship between scores produces a correlation close to 0. Consistency of scores, or the correlation of a test with itself, is believed to indicate that a score is reliable under the criteria established within Classical Test Theory [Hack #6]. Classical Test Theory suggests, among other things, that random error is the only reason that scores for a single person will vary if the same test is taken many times.

HACK #32   Establish Validity

The single most important characteristic of a test is that it is useful for its intended purpose. Establishing validity is important if anyone is to trust that a test score means what it is supposed to mean. You can convince yourself and others that your test is valid if you provide certain types of evidence.

A good test measures what it is intended to measure. For example, a survey that is supposed to find out how often high school students wear seatbelts should, obviously, contain questions about seatbelt use. A survey without these items could reasonably be criticized as not having *validity*. Validity is the extent to which something measures whatever it is expected to measure. Surveys, tests, and experiments all require validity to be acceptable. If you are building a test for psychological or educational measurement, or just want to be sure your test is useful, you should be concerned about establishing validity.

Validity is not something that a test score either has or does not have. Validity is an argument that is made by the test designer, those relying on the test's results, or anyone else who has a stake in the acceptance of the test and its results.

Consider a spelling test that consists of math problems. Clearly, a test with math problems is not a valid spelling test. While it is not a valid spelling test, though, it might well be a valid math test. The validity of a test or survey is not in the instrument itself, but in the interpretation of the results.

A test might be valid for one purpose, but not another. It would not be appropriate to interpret a child's score on a spelling test as an indication of her math ability; the score might be valid as a measure of verbal ability, but not as a measure of numerical fluidity. The score itself is neither valid nor invalid; it is the meaning attached to the score that is arguably valid or not valid.

To illustrate how to solve the problem of establishing validity, imagine you have designed a new way of measuring spelling ability. You want to sell the test forms to school districts across the country, but first you must produce visible evidence that your test measures spelling ability and not something else, such as vocabulary, test anxiety, reading ability, or (in terms of other factors that might affect scores) gender or race.

Strategies for Winning the Validity Argument

Validity might seem like an argument that can never be won, because as an invisible indicator of quality, it can never be completely established. As a test developer, though, you want to be able to convince your test-takers and anyone who will be using the results of your test that you are measuring substantially whatever it is you are supposed to measure. Fortunately, there are a number of accepted ways in which evidence for the validity of a test can be provided.

The most commonly accepted type of validity evidence is also, interestingly, theoretically the weakest argument one can make for validity. This argument is one of *face validity*, and it runs as follows: this test is valid because it looks (on its face) like it measures what it is supposed to measure. Those presenting or accepting an argument for face validity believe that the test in question has the sort of items that one would expect to find on such a test. For example, the seatbelt use survey mentioned earlier would be accepted as valid if it has items asking about seatbelt use.

The face validity argument is weak because it relies on human judgment alone, but it can be compelling. Common sense is a strong argument, perhaps even the strongest, for convincing someone to accept any aspect of an assessment. Though face validity seems less scientific than other types of validity evidence (and in a real sense, it *is* less scientific), few test instruments would be acceptable to those who make and use them if face validity evidence is lacking. If you, as a test developer or user, cannot supply the types of validity evidence discussed in the rest of this hack, you are expected to provide a test that at least has face validity.

> For your spelling test, if test takers are asked to spell, you have established face validity.

Four somewhat more scientific types of validity evidence are generally accepted by those who rely on assessments. They are all part of the range of arguments that can be made for validity.

Content-based arguments
> Do the items on the test fairly represent the items that could be on the test? If a test is meant to cover some well-defined domain of knowledge, do the questions fairly sample from that domain?

Criterion-based arguments
> Do scores on the test estimate performance on some other test?

Construct-based arguments
> Does the score on the test represent the trait or characteristic you wish to measure?

Consequences-based arguments
> Do the people who take the test benefit from the experience? Is the test biased against certain groups? Does taking the test cause so much stress that, no matter how you score, it isn't worth it?

Content-Based Arguments

If you decide to measure a concept, there are many aspects of that concept and many different questions that can be asked on a test. Some demonstration that the items you choose for your test represent all possible items would be a content-based argument for validity.

This sounds like a daunting requirement. Traditionally, this sort of evidence has been considered more important for tests of achievement. In areas of achievement—medicine, law, English, mathematics—there are fairly well-defined domains and content areas from which a valid test should sample. A classroom teacher also, presumably, has defined a set of objectives or content areas that a test should measure. Such concisely defined aspects of a subject are rarely available, however, when testing a range of behaviors, knowledge, or attitudes. Consequently, making a reasonable argument that you have selected questions that are representative of some imaginary pool of all possible questions is difficult.

So, what is necessary for content evidence of validity in test construction? It seems that, at a minimum, test construction calls for some organized method of question selection or construction. When measuring self-esteem, for example, questions might cover how the test taker feels about himself in different environments (e.g., work, home, or school), while performing different tasks (e.g., sports, academics, or job duties), or how he feels about different aspects of himself (e.g., his appearance, intelligence, or social skills).

For a classroom teacher measuring how much students have learned during the last few weeks, a *table of specifications* (an organized list of topics covered and weights indicating their importance) is a good method.

The choice of how to organize a concept or how to break it down into components belongs to the test developer. The developer might have been inspired by research or other tests, or she might just be following a common-sense scheme. The key is to convince yourself, so that you can convince others that you are covering the vital aspects of whatever area you are measuring.

For your spelling test, if you can establish that the words students are asked to spell represent a larger pool of words that students should be able to spell, you are providing content-based validity evidence.

Criterion-Based Arguments

Criterion evidence of validity demonstrates that responses on a test predict performance in some other situation. "Performance" can mean success in a job, a test score, ratings by others, and so on.

If responses on the test are related to performance on criteria that can be measured immediately, the validity evidence is referred to as *concurrent validity*. If responses on the test are related to performance on criteria that cannot be measured until some future time (e.g., eventual college graduation, treatment success, or eventual drug abuse), the validity evidence is called *predictive validity*.

It might go without saying that the measures you choose to support criterion validity should be relevant; the criteria should be measures of concepts that are somehow theoretically related. This form of validity evidence is most persuasive and important when the express purpose of a test is to estimate or predict performance on some other measure.

Criterion-based evidence is less persuasive, and perhaps irrelevant, for tests that do not claim to predict the future or estimate performance on some other measure. For example, such evidence might not be useful for your spelling test. On the other hand, it is possible that you can demonstrate that high scorers on your test do well in the National Spelling Bee.

Construct-Based Arguments

The third category of validity evidence is construct evidence. A *construct* (pronounced with an emphasis on the first syllable: *con*-struct) is the theoretical concept or trait that a test is designed to measure. We know that we can never measure constructs such as intelligence or self-esteem directly. The methods of psychological measurement are indirect. We ask a series of questions we hope will require the respondent to use the part of her mind we are measuring or reference the portion of her memory that contains information on past behaviors or knowledge, or, at the very least, direct the respondent to examine her attitudes and feelings on a particular topic.

We further hope that the test takers accurately and honestly respond to test items. In practice, test results are often treated as a direct measure of a construct, but we shouldn't forget that they are educated guesses only. The success of this whole process depends on another set of assumptions: that we have correctly defined the construct we are trying to measure and that our test mirrors that definition.

Construct evidence, then, often includes both a defense of the defined construct itself and a claim that the instrument used reflects that definition. Evidence presented for construct validity can include a demonstration that responses behave as theory would expect responses to behave. Construct validity evidence continues to accumulate whenever a survey or test is used, and, like all validity arguments, it can never be fully convincing. In a sense, construct validity arguments include both content and criterion validity arguments, because all validity evidence seeks to establish a link between a concept and the activity that claims to measure it.

For your spelling test, there might be research on the nature of *spelling ability* as a cognitive activity or personality trait or some other well-defined entity. If you can define what you mean by spelling ability and demonstrate that your test's scores behave as your definition would expect, then you can claim construct-based validity evidence. Does theory suggest that better readers are better spellers? Show that relationship, perhaps with a correlation coefficient [Hack #11], and you have presented validity evidence that might convince others.

Consequences-Based Arguments

Until the last decade or two, measurement folks interested in establishing validity were concerned only with demonstrating that the test score reflected the construct. Because of increasing concerns that certain tests might unfairly penalize whole groups of people, plus other concerns about the social consequences of the common use of tests, policy makers and measurement philosophers now look at the consequences experienced by the test taker because of taking a test.

The idea is that we have gotten so used to testing and making high-stakes decisions based on those test scores that we should take a step back occasionally and ask whether society is really better off if we rely on tests to make these decisions. This represents a broadening of the definition of validity from *a score representing the construct* to *a test fulfilling its intended purpose*. Presumably, tests are here to help the world, not hurt it, and consequences-based validity evidence helps to demonstrate the societal value of testing.

> Like people from the government in all those old jokes, tests are "here to help us."

For your spelling test, the key negative consequences you want to rule out involve test bias. If your theory of spelling ability expects no differences across gender, race, or socio-economic status, then spelling scores should be equal between those groups. Produce evidence of similar scores between groups, perhaps with a *t* test [Hack #17], and you will be well on your way to establishing that your test is fair and valid.

Choosing from the Menu of Validity Options

The variety of categories of validity evidence described here represents a strategic menu of options. If you want to demonstrate validity, you can choose from across the range of validity evidence types.

Clearly, not all tests need to provide all types of validity evidence. A small teacher-made history test meant for a group of 25 students might require only some content-based validity evidence to convince the teacher to trust the results. Criterion-based validity evidence is unnecessary, because estimating performance on another test is not an intended purpose of this sort of test.

On the other hand, higher-stakes tests, such as college admissions tests (e.g., the ACT, SAT, and GRE) and intelligence tests used to identify students as eligible for special education funding, should be supported with evidence from all four validity areas. For your spelling test, you can decide which type of evidence, and which type of argument, is most convincing.

Predict the Length of a Lifetime
HACK #33

Many of us instinctively trust that things that have been around a long time are likely to be around a lot longer, and things that haven't, aren't. The formalization of this heuristic is known as Gott's Principle, and the math is easy to do.

Physicist J. Richard Gott III has so far correctly predicted when the Berlin Wall would fall and calculated the duration of 44 Broadway shows.[1] Controversially, he has predicted that the human race will probably exist between 5,100 and 7.8 million more years, but no longer. He argues that this is a good reason to create self-sustaining space colonies: if the human race puts some eggs in other nests, we might extend the life span of our species in case of an asteroid strike or nuclear war on the home planet.[2]

Gott believes that his simple calculations can be extended to almost anything at all, within certain parameters. To predict how long something will be around by using these calculations, all you need to know is how long it *has* been around already.

In Action

Gott bases his calculations on what he calls the Copernican Principle (and what some people call, in this specific application, Gott's Principle). The principle says that when you choose a moment in time to calculate the lifetime of a phenomenon, that moment is probably quite ordinary, not special or privileged, just as Copernicus told us the Earth does not occupy a privileged place in the universe.

It's important to choose subjects at ordinary, unprivileged moments. Biasing your test by choosing subjects that you already believe to be near the beginning or end of their life span—such as the human occupants of a neonatal ward or a nursing home—will yield bad results. Further, Gott's Principle is less useful in situations where actuarial data already exists. Plenty of actuarial data is available on the human life span already, so Gott's Principle is less useful here.

Having chosen a moment, let's examine it. All else being equal, there's a 50 percent chance the moment is somewhere in the middle 50 percent of the phenomenon's lifetime, a 60 percent chance it's in the middle 60 percent, a 95 percent chance it's in the middle 95 percent, and so on. Therefore, there's only a 25 percent chance that you've chosen a moment in the first fourth of its lifetime, a 20 percent chance it's in the first fifth, a 2.5 percent chance it's in the last 2.5 percent of the subject's lifetime, and so on.

Table 3-11 provides equations for the 50 percent, 60 percent, and 95 percent confidence levels. The variable tpast represents how long the object has existed, and tfuture represents how long it is expected to continue.

Table 3-11. Confidence levels under Gott's Principle

Confidence level	Minimum tfuture	Maximum tfuture
50 percent	$t_{past}/3$	$3t_{past}$
60 percent	$t_{past}/4$	$4t_{past}$
95 percent	$t_{past}/39$	$39t_{past}$

Let's look at a simple example. Quick: whose work do you think is more likely to be listened to 50 years from now, Johann Sebastian Bach's or Britney Spears'? Bach's first work was performed around 1705. At the time of this writing, that's 300 years ago. Britney Spears' first album was released in January 1999, about 6.5 years or 79 months ago.

Consulting Table 3-11, for the 60 percent confidence level, we see that the minimum t_{future} is $t_{past}/4$, and the maximum is $4t_{past}$. Since t_{past} for Britney's music is 79 months, there is a 60 percent chance that Britney's music will be heard for between 79/4 months and 79×4 months longer. In other words, we can be 60 percent sure that Britney will be a cultural force for somewhere between 19.75 months (1.6 years) and 316 months (26.3 years) from now.

Sixty percent is a good confidence level for quick estimation; not only is it a better-than-even chance, but the factors 1/4 and 4 are easy to use.

By the same token, we can expect people to listen to Bach's music for somewhere between another 300/4 and 300×4 years at the 60 percent confidence level, or somewhere between 75 years and 1,200 years from now. Thus, we can predict that there's a good chance that Britney's music will die with her fans, and there's a good chance that Bach will be listened to in the fourth millennium.

How It Works

Suppose we are studying the lifetime of some object that we'll call the *target*. As we've already seen, there's a 60 percent chance we are somewhere in the middle 60 percent of the object's lifetime (Figure 3-4).[3]

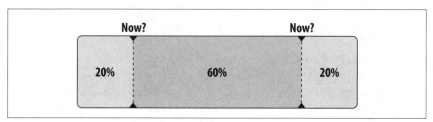

Figure 3-4. The middle 60 percent of the lifetime

If we are at the very end of this middle 60 percent, we are at the second point marked "now?" in Figure 3-4. At this point, only 20 percent of the target's lifetime is remaining (Figure 3-5), which means that t_{future} is equal to one-fourth of t_{past} (80 percent). This is the minimum remaining lifetime we expect at the 60 percent confidence level.

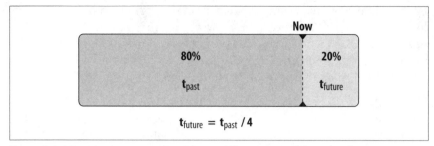

Figure 3-5. *The minimum remaining lifetime (60 percent confidence level)*

Similarly, if we are at the beginning of the middle 60 percent (the first point marked "now?" in Figure 3-4), 80 percent of the target's existence lies in the future, as depicted in Figure 3-6. Therefore, t_{future} (80 percent) is equal to 4 $\times t_{past}$ (20 percent). This is the maximum remaining lifetime we expect at the current confidence level.

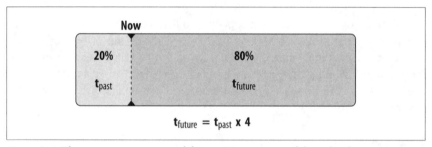

Figure 3-6. *The maximum remaining lifetime (60 percent confidence level)*

Since there's a 60 percent chance we're between these two points, we can calculate with 60 percent confidence that the future duration of the target (t_{future}) is between $t_{past}/4$ and $4 \times t_{past}$.

In Real Life

Suppose you want to invest in a company and you want to estimate how long the company will be around to determine whether it's a good investment. You can use Gott's Principle to do so. Although it's not publicly traded, let's take O'Reilly Media, the publisher of this book, as an example.

I certainly didn't pick O'Reilly Media at random, and plenty of historical information is available about how long companies tend to last, but let's try Gott's Principle as a rough-and-ready estimate of O'Reilly's longevity anyway. After all, there's probably good data on the longevity of Broadway shows, but Gott didn't shrink from analyzing them—and I hesitate to say that now that O'Reilly has published *Mind Performance Hacks*, its immortality is assured.

According to the Wikipedia, O'Reilly started in 1978 as a consulting firm doing technical writing. It's July 2005 as I write this, so O'Reilly has existed as a company for approximately 27 years. How long can we expect O'Reilly to continue to exist?

Here's O'Reilly's likely lifetime, calculated at the 50 percent confidence level:

Minimum
   27/3 = 9 years (until July 2014)

Maximum
   27×3 = 81 years (until July 2086)

Here are our expectations at the 60 percent confidence level:

Minimum
   27/4 = 6 years and 9 months (until April 2012)

Maximum
   27×4 = 108 years (until July 2113)

Finally, here's our prediction with 95 percent confidence:

Minimum
   27/39 = 0.69 years = about 8 months and 1 week (until mid-March 2006)

Maximum
   27×39 = 1,053 years (until July 3058)

In the post-dot-com economy, these figures look pretty good. For example, Apple Computer's aren't much better, and Microsoft was founded in 1975, so the same can be said for it. A real investor would want to consider many other factors, such as annual revenue and stock price, but as a first cut, it looks as though O'Reilly Media is at least as likely to outlive a hypothetical investor as to tank in the next decade.

Endnotes

1. Ferris, Timothy. "How to Predict Everything." *The New Yorker*, July 12, 1999.

2. Gott, J. Richard III. "Implications of the Copernican Principle for Our Future Prospects." *Nature*, 363, May 27, 1993.

3. Gott, J. Richard III. "A Grim Reckoning." *http://pthbb.org/manual/services/grim*.

—Ron Hale-Evans

HACK #34 Make Wise Medical Decisions

Medical tests provide diagnostic screening information that is often misunderstood by patients and, sometimes, even by doctors. Understanding the probability characteristics called "sensitivity" and "specificity" can provide a more accurate and (sometimes) reassuring picture.

As a consumer of medical information, you have to make decisions about behavior, treatment, seeking a second opinion, and so on. You likely rely on medical information—newspaper stories, your doctor's advice, test results—to make those decisions. However, much of the medical information you get from your doctor has a known amount of error. This is especially true about diagnostic test results that indicate the probability that you have a certain condition.

This hack is all about using information about the characteristics of those medical tests to get a more accurate picture of reality and, hopefully, make better decisions about treatment.

Statistics and Medical Screening

To use medical test information wisely, we have to learn just a bit about what the concept of *accuracy* means for these tests. The four possible outcomes of medical tests, in terms of accuracy, are shown in Table 3-12.

Table 3-12. Possible medical test outcomes

	Patient actually has the condition (A)	Patient actually does not have condition (B)
Test result indicates patient has condition	True positive (score is correct)	False positive (score is wrong)
Test result indicates patient does not have condition	False negative (score is wrong)	True negative (score is correct)

The reliability [Hack #6] of medical screening tests is summarized by two proportions called *sensitivity* and *specificity*. Essentially, those who rely on these tests are concerned with three questions of accuracy:

- If a person has the disease, how likely is the person to score a positive test result? This likelihood is *sensitivity*. Of those people in column A, what percent will receive a positive test result?

- If the person does *not* have the disease, how likely is the person to score a negative test result? This likelihood is *specificity*. Of those people in column B, what percent will receive a negative test result?

- If a person scores a positive test result, how likely is the person to have the disease? From the patient's perspective, this is the ultimate question, and it can be thought of as the basic validity concern with these tests. Doctor, can I trust these test results, or could there be some mistake?

Notice in Table 3-12 that there are different people in columns A and B. People with the disease are in column A and people without the disease are in column B. If you are in column A, you cannot score a false positive on the test, because a positive result is correct. If you are in column B, you cannot score a false negative, because a negative result is correct.

Which column anyone is in depends on the natural distribution of the disease. The chance that someone will be in column A (the chance the person actually has the disease) depends on the *base rate* of the disease. If 5 percent of the population has the disease, 5 percent of the population would find themselves in column A.

Understanding Breast Cancer Screening

Breast cancer is an example of a serious condition for which there are diagnostic screening tests. Breast cancer screening begins with a mammogram test. A positive result on this test results in further testing: another mammogram, ultrasound, or biopsy.

We are first interested in answering the questions regarding the sensitivity and specificity of breast cancer screening. With that information and knowledge of the base rate for breast cancer, we can answer the most important question:

> If a woman scores a positive test result, how likely is she to have breast cancer?

By asking your doctor or doing some research, you might discover that sensitivity for mammograms is about 90 percent. Specificity is about 92 percent.

The exact sensitivity and specificity for breast cancer screenings change over time as different populations take the test. Younger women now have mammograms more commonly than in the past, and the test is less sensitive and less specific for younger women. Of course, you should check with a physician or expert for current levels of precision.

Table 3-13 shows those numbers in the layout used in Table 3-12. Because columns A and B must both independently add to 100 percent, we can also estimate the rate of false negatives and rate of false positives.

Table 3-13. Theoretical mammogram results for 10,000 women

	Patient actually has breast cancer (A) N=120	Patient actually does not have breast cancer (B) N=9,880
Mammogram indicates cancer	Sensitivity 90 percent N=108	False positives 8 percent N=790
Mammogram does not indicate cancer	False negatives 10 percent N=12	Specificity 92 percent N=9,090

Table 3-13 also shows the outcomes for 10,000 hypothetical women, based on the base rate of breast cancer in the population, which is about 1.2 percent.

It turns out that it is difficult to identify an accurate incident rate for breast cancer because of the different ways one can define the relevant population and, of course, limitations in the accuracy of breast cancer testing. I'm using an often-reported and fairly well-accepted estimate of the current percentage of women aged 40 to 84 that have breast cancer.

Let's return now to the third question in our list of important questions to ask before interpreting the results of a medical test. If a person scores a positive test result, how likely is the person to have the disease? Out of 10,000 women who have a breast cancer screening, 898 will receive a positive score. For 790 of those women, the score is wrong; they do not actually have breast cancer. For 108 of those women, the test was right; they do have cancer. In other words, if a person scores a positive result, it is only 12 percent likely that they have the disease. The most common result for follow-up testing to a positive mammogram is that the patient is, in fact, cancer free.

What about the accuracy of a negative result? Of the 9,102 women who will score negative on the screening, 12 actually have cancer. This is a relatively small 1/10 of 1 percent, but the testing will miss those people altogether, and they will not receive treatment.

Why It Works

Medical screening accuracy uses a specific application of a generalized approach to conditional probability attributed to Thomas Bayes, a philosopher and mathematician in the 1700s. "If this, then what are the chances that..." is a conditional probability question.

Bayes's approach to conditional probabilities was to look at the naturally occurring frequencies of events. The basic formula for estimating the chance that one has a disease if one has a positive test result is:

$$\frac{\text{True Positives}}{\text{True Positives} + \text{False Positives}}$$

Expressed as conditional probabilities, the formula is:

$$\frac{\text{Base Rate} \times \text{Sensitivity}}{(\text{Base Rate} \times \text{Sensitivity}) + (1 - \text{Base Rate})(1 - \text{Specificity})}$$

To answer the all-important question in our breast cancer example ("If a woman scores a positive test result, how likely is she to have breast cancer?"), the mammogram equation takes on these values:

$$\frac{.012 \times .90}{(.012 \times .90) + (1 - .012)(1 - .92)} = .1202$$

Making Informed Decisions

Medical tests are used to indicate whether patients might have a disease or be at risk for getting one. Identifying the presence or absence of a disease such as cancer is a process that usually has at least two steps. In step one, a patient is administered a screening test, typically a relatively simple and non-invasive test that looks for indications that a person might have a certain medical condition. If the result is positive, the second step is to conduct a second test (or series of tests) that is typically more complex, invasive, and expensive, but also much more accurate, to confirm or disconfirm the original finding.

Medical tests are not perfectly reliable and valid. Test results can be wrong. There are four possibilities for anyone who undergoes medical testing. A patient might have the disease and the test indicates this, or the patient does not have the disease and the test finds no presence of it. In these cases, the test worked right and the scores are valid.

Conversely, the test results might reflect the opposite of the true medical condition, with a positive result wrongly indicating presence of a disease that is not there, or a negative result wrongly indicating that the patient is disease-free. In these cases, the test did not work right and the results are not valid. This table of outcomes is similar to the possibilities when one accepts or rejects a hypothesis in statistical decision making [Hack #4].

Breast cancer screening is very good at finding breast cancer when it is there to find. However, one drawback to such a sensitive test for a low-incidence disease is that many more people will be told that they might have the disease than actually do. There is a trade-off in medical testing between test sensitivity and test specificity. More sensitive tests tend to result in more false positives, but in serious situations like life and death, this seems to be a result we can live with.

See Also

- Gigerenzer, G. (2002). *Calculated risks. How to know when numbers deceive you.* New York: Simon and Schuster.

Beating the Odds
Hacks 35–49

Why risk more than you have to when you take risks? Casino games require you to take some chances, but this chapter of real-world stat hacks will help you keep your edge and perhaps even overcome the house's edge.

Start with Texas Hold 'Em poker [Hack #36]. (Maybe you've heard of it?) When you play poker [Hack #37], play the odds [Hack #38].

Make sure, of course, to always gamble smart [Hack #35], regardless of what you play, though when it comes to the level of risk you take, some games [Hacks #39 and #40] are better than others [Hack #41].

If you like to make friendly wagers with friends or strange wagers with strangers, you can use the power of statistics to win some surprisingly winnable bar bets with cards [Hacks #42 and #44], dice [Hack #43], or just about anything else you can think of [Hack #46], including your friends' birthdays [Hack #45].

Speaking of weird gambling games (and I think we were), there are some odd statistical quirks [Hacks #47 and #49] you'll need to know when you play them, even if it is just flipping a coin [Hack #48].

HACK #35 Gamble Smart

Whatever the game, if money and chance are involved, there are some basic gambling truths that can help the happy statistician stay happy.

Although this chapter is full of hacks aimed at particular games, many of them games of chance, there are a variety of tips and tools that are useful across the board for all gamblers. Much mystery, superstition, and mathematical confusion pervade the world of gambling, and knowing a little more about the geography of this world should help you get around. This hack shows how to gamble smarter by teaching you about the following things:

- The Gambler's Fallacy, an intuitive yet false belief system that has cost many an otherwise well-informed gamer
- Casinos and money
- Systems, sophisticated money management, and wagering procedures that do not work

The Gambler's Fallacy

Did you ever have so many bad blackjack hands in a row that you increased your bet, knowing that things were due to change anytime now? If so, you succumbed to the *gambler's fallacy*, a belief that because there are certain probabilities expected in the long run, a short-term streak of bad luck is likely to change soon.

The gambler's fallacy is that there is a swinging pendulum of chance and it swings in the region of bad outcomes for a while, loses momentum, and swings back into a region of good outcomes for a while. The problem with following this mindset is that luck, as it applies to games of pure chance, is a series of independent events, with each individual outcome unrelated to the outcome that came before. In other words, the location of the pendulum in a *good* region or *bad* region is unrelated to where it was a second before, and—here's the rub—there isn't even a pendulum. The fickle finger of fate pops randomly from possible outcome to possible outcome, and the probability of it appearing at any outcome is the probability associated with each outcome. There is no momentum. This truth is often summarized as "the dice have no memory."

Examples of beliefs consistent with the gambler's fallacy include:

- A slot machine that hasn't paid out in a while is due.
- A poker player who has had nothing but bad hands all evening will soon get a super colossal hand to even things out.
- A losing baseball team that has lost the last three games is more likely to win the fourth.
- Because rolling dice and getting three 7s in a row is unlikely to occur, rolling a fourth after having just rolled three straight must be basically impossible.
- A roulette ball that has landed on eight red numbers in a row pretty much must hit a black number next.

Avoid fallacies like this at all costs, and gambling should cost you less.

Casinos and Money

Casinos make money. One reason they make a profit is that the games themselves pay off amounts of money that are slightly less than the amount of money that would be fair. In a game of chance, a *fair* payout is one that makes both participants, the casino and the player, break even in the long run.

An example of a fair payout would be for casinos to use roulette wheels with only 36 numbers on them, half red and half black. The casino would then double the money of those who bet on red after a red number hits. Half the time the casino would win, and half the time the player would win. In reality, American casinos use 38 numbers, two of them neither red nor black. This gives the house a 2/38 edge over a fair payout. Of course, it's not unfair in the general sense for a casino to make a profit this way; it's expected and part of the social contract that gamblers have with the casinos. The truth is, though, that if casinos made money only because of this edge, few would remain in business.

The second reason that casinos make money is that gamblers do not have infinitely deep pockets, and they do not gamble an infinite period of time. The edge that a casino has—the 5.26 percent on roulette, for example—is only the amount of money they would take if a gambler bet an infinite number of times. This infinite gambler would be up for a while, down for a while, and at any given time, on average, would be down 5.26 percent from her starting bankroll.

What happens in real life, though, is that most players stop playing sometime, usually when they are out of chips. Most players keep betting when they have money and stop betting when they don't. Some players, of course, walk away when they are ahead. No player, though, keeps playing when they have no money (and no credit).

Imagine that Table 4-1 represents 1,000 players of any casino game. All players started with $100 and planned to spend an evening (four hours) playing the games. We'll assume a house edge of 5.26 percent, as roulette has, though other games have higher or lower edges.

Table 4-1. Fate of 1,000 hypothetical gamblers

Time spent playing	Have some money left	Mean bankroll left	Have lost all their money	Still playing
After an hour of play	900	$94.74	100	900
After two hours of play	800	$94.74	200	800

Table 4-1. Fate of 1,000 hypothetical gamblers (continued)

Time spent playing	Have some money left	Mean bankroll left	Have lost all their money	Still playing
After three hours of play	700	$94.74	300	700
After four hours of play	600	$94.74	400	600

In this example—which uses made-up but, I bet, conservative data—after four hours, the players still have $56,844, the casino has $43,156, and from the total amount of money available, the casino took 43.16 percent. That's somewhat more than the official 5.26 percent *house edge*.

It is human behavior—the tendency of players to keep playing—not the probabilities associated with a particular game, that makes gambling so profitable for casinos. Because the house rules are published and reported, statisticians can figure the house edge for any particular game.

Casinos are not required to report the actual money they take in from table games, however. Based on the depth of the shag carpet at Lum's Travel Inn of Laughlin, Nevada (my favorite casino), though, I'm guessing casinos do okay. The general gambler's hack here is to walk away after a certain period of time, whether you are ahead or behind. If you are lucky enough to get far ahead before your time runs out, consider running out of the casino.

Systems

There are several general betting systems based on money management and changing the amount of your standard wager. The typical system suggests increasing your bet after a loss, though some systems suggest increasing your bet after a win. As all these systems assume that a streak, hot or cold, is always more likely to end than continue, they are somewhat based on the gambler's fallacy. Even when such systems make sense mathematically, though, anytime wagers must increase until the player wins, the *law of finite pocket size* sabotages the system in the long run.

Here's a true story. On my first visit to a legal gambling establishment as a young adult, I was eager to use a system of my own devising. I noticed that if I bet on a column of 12 numbers at roulette, I would be paid 2 to 1. That is, if I bet $10 and won, I would get my $10 back, plus another $20. Of course, the odds were against any of my 12 numbers coming up, but if I bet on *two* sets of 12 numbers, then the odds were with me. I had a 24 out of 36 (okay, really 38) chance of winning—better than 50 percent!

I understood, of course, that I wouldn't triple my money by betting on two sets of numbers. After all, I would lose half my wager on the set of 12 that didn't come up. I saw that if I wagered $20, about two-thirds of the time I would win back $30. That would be a $10 profit. Furthermore, if I didn't win on the first spin of the wheel, I would bet on the same numbers again, but this time I would double my bets! (I am a super genius, you agree?) If by some slim chance I lost on that spin as well, I would double my bet one *more* time, and then win all my money back, plus make that 50 percent profit. To make a long story short, I did just as I planned, lost on all three spins and had no money left for the rest of the long weekend and the 22-hour drive home.

The simplest form of this sort of system is to double your bet after each loss, and then whenever you do win (which you are bound to do), you are back up a little bit. The problem is that it is typical for a long series of losses to happen in a row; these are the normal fluctuations of chance. During those losing streaks, the constant doubling quickly eats up your bankroll.

Table 4-2 shows the results of doubling after just six losses in a row, which can happen frequently in blackjack, roulette, craps, video poker, and so on.

Table 4-2. The "double after a loss" system

Loss number	Bet size	Total expenditure
1	$5	$5
2	$10	$15
3	$20	$35
4	$40	$75
5	$80	$155
6	$160	$315

Six losses in a row, even under an almost 50/50 game such as betting on a color in roulette, is very likely to happen to you if you play for more than just a couple of hours. The actual chance of a loss on this bet for one trial is 52.6 percent (20 losing outcomes divided by 38 possible outcomes). For any six spins in a row, a player will lose all spins 2.11 percent of the time (.526 × .526 × .526 × .526 × .526 × .526).

Imagine 100 spins in two hours of play. A player can expect six losses in a row to occur twice during that time. Commonly, then, under this system, a player is forced to wager 32 times the original bet, just to win an amount equal to that original bet. Of course, most of the time (52.6 percent), when there have been six losses in a row, there is then a seventh loss in a row!

Systems do exist for gambling games in which players can make informed strategic decisions, such as blackjack (with card counting) and poker (reading your opponent), but in games of pure chance, statisticians have learned to expect the expected.

Know When to Hold 'Em

HACK
#36

In Texas Hold 'Em, the "rule of four" uses simple counting to estimate the chance that you are going to win all those chips.

Texas Hold 'Em No Limit Poker is everywhere. As I write this, I could point my satellite dish to ESPN, ESPN2, ESPN Classics, FOX Sports, Bravo, or E! and see professional poker players, lucky amateurs, major celebrities, minor celebrities, and even (Lord help us, on the Speed channel) NASCAR drivers playing this simple game.

You probably play yourself, or at least watch. The most popular version of the game is simple. All players start with the same amount of chips. When their chips are gone, so are they. Every round, players get two cards each that only they (and the patented tiny little poker table cameras) see. Then, three community cards are dealt face up. This is the *flop*. Another community card is then dealt face up. That's the *turn*. Finally, one more community card, the *river*, is dealt face up. Betting occurs at each stage. Players use any five of the seven cards (five community cards, plus the two they have in their hands) to make the best five-card poker hand they can. The best hand wins.

Because some cards are face up, players have information. They also know which cards they have in their own hands, which is more information. They also know the distribution of all cards in a standard 52-card deck. All this information about a known distribution of values [Hack #1] makes Texas Hold 'Em a good opportunity to stat hack all over the place [Hacks #36 and #38].

One particularly crucial decision point is the round of betting right after the flop. There are two more cards to come that might or might not improve your hand. If you don't already have *the nuts* (the best possible hand), it would be nice to know what the chances are that you will improve your hand on the next two cards. The *rule of four* allows you to easily and fairly accurately estimate those chances.

How It Works

The rule of four works like this. Count the number of cards (without moving your lips) that could come off of the deck that would help your hand. Multiply that number by four. That product will be the percent chance that you will get one or more of those cards.

Example 1. You have a Jack of Diamonds and a Three of Diamonds. The flop comes King of Clubs, Six of Diamonds, and Ten of Diamonds. You have four cards toward a flush, and there are nine cards that would give you that flush. Other cards could help you, certainly (a Jack would give you a pair of Jacks, for example), but not in a way that would make you feel good about your chances of winning.

So, nine cards will help you. The rule of four estimates that you have a 36 percent chance of making that flush on either the turn or the river ($9 \times 4 = 36$). So, you have about a one out of three chance. If you can keep playing without risking too much of your stack, you should probably stay in the hand.

Example 2. You have an Ace of Diamonds and a Two of Clubs. The flop brings the King of Hearts, the Four of Spades, and the Seven of Diamonds. You could count six cards that would help you: any of the three Aces or any of the three Twos. A pair of twos would likely just mean trouble if you bet until the end, so let's say there are three cards, the Aces, that you hope to see. You have just a 12 percent chance ($3 \times 4 = 12$). Fold 'em.

Why It Works

The math involved here rounds off some important values to make the rule simple. The thinking goes like this. There are about 50 cards left in the deck. (More precisely, there are 47 cards that you haven't seen). When drawing any one card, your chances of drawing the card you want **[Hack #3]** is that number divided by 50.

> I know, it's really 1 out of 47. But I told you some things have been simplified to make for the simple mnemonic "the rule of four."

Whatever that probability is, the thinking goes, it should be doubled because you are drawing twice.

> This also isn't quite right, because on the river the pool of cards to draw from is slightly smaller, so your chances are slightly better.

For the first example, the rule of four estimates a 36 percent chance of making that flush. The actual probability is 35 percent. In fact, the estimated and actual percent chance using the rule of four tends to differ by a couple percentage points in either direction.

Other Places It Works

Notice that this method also works with just one card left to go, but in that case, the rule would be called the *rule of two*. Add up the cards you want and multiply by two to get a fairly accurate estimate of your chances with just the river remaining. This estimate will be off by about two percentage points in most cases, so statistically savvy poker players call this the *rule of two plus two*.

Where It Doesn't Work

The rule of four will be off by quite a bit as the number of cards that will help you increases. It is fairly accurate with 12 *outs* (cards that will help), where the actual chance of drawing one of those cards is 45 percent and the rule of four estimate is 48 percent, but the rule starts to overestimate quite a bit when you have more than 12 cards that can help your hand.

To prove this to yourself without doing the calculations, imagine that there are 25 cards (out of 47) that could help you. That's a great spot to be in (and right now I can't think of a scenario that would produce so many outs), but the rule of four says that you have a 100 percent chance of drawing one of those cards. You know that's not right. After all, there are 22 cards you could draw that don't help you at all. The real chance is 79 percent. Of course, making a miscalculation in this situation is unlikely to hurt you. Under either estimate, you'd be nuts to fold.

HACK #37   Know When to Fold 'Em

In Texas Hold 'Em, the concept of pot odds provides a powerful tool for deciding whether to call or fold.

If you watch any poker on TV, you quickly pick up a boatload of jargon. You'll hear about *big slick* and *bullets* and *all-in* and *tilt*. You'll also hear discussions about *pot odds*, as in, "He might call here, not because he thinks he has the best hand, but because of the pot odds."

When the pot odds are right, you should call a hand even when the odds are that you will lose. So, what are pot odds and why would I ever put more money into a pot that I am likely to lose?

Pot Odds

Pot odds are determined by comparing the chance that you will win the pot to the amount of chips you would win if you did win the pot. For example, if you estimate that there is a 50 percent chance that you will win a pot, but the pot is big enough that winning it would win you more than double the cost of calling the bet in front of you, then you should call.

To see how pot odds works in practice, here is a scenario with four players: Thelma, Louise, Mike, and Vince. As shown in Table 4-3, Thelma is in the best shape before the flop.

> The tables that follow show the decisions each player makes based on the pot odds at each point in a round. Read the following tables left to right, following each column all the way down, to see what Thelma thinks and does, then what Louise thinks and does, and so on.

Table 4-3. Players' starting hands

Player	Thelma	Louise	Mike	Vince
Cards	Ace Clubs, Ace Hearts	2 Clubs, 4 Clubs	4 Hearts, 5 Spades	King Diamonds, 10 Diamonds
Opening bet	50	50	50	50

Then comes the flop: Ace Spades, 3 Diamonds, 6 Diamonds. Table 4-4 shows the revised analysis of the players' positions. After the flop, three of them are hoping to improve their hands, while one of them, Thelma, would be satisfied with no improvement of her hand, thinking she has the best one now. Thelma is driving the betting, and the other three players are deciding whether to call.

Table 4-4. Analysis after the flop

Player	Thelma	Louise	Mike	Vince
Needed cards		Any of four 5s	Any of four 2s or four 7s	Any of nine diamonds
Chance of getting card		16 percent	32 percent	36 percent
Current pot	200	250	250	300
Cost to call as percentage of pot		20 percent	20 percent	17 percent
Action	Bet 50	Fold	Call 50	Call 50

Table 4-4 shows the use of *pot odds* after the flop. Thelma has a pair of aces to start and hits the third ace on the flop. Consequently, she begins each round by betting. The other players who have yet to hit anything must decide whether to stick around and hope to improve their hands into strong, likely winners.

Pot odds come into play primarily when making the decision whether to stick around or fold. Louise needs a five to make her straight, and she estimates a 16 percent chance of getting that 5 somewhere in the next two cards. However, with that pot currently at $250 and a $50 raise from Thelma, which she would have to call, Louise would have to pay 20 percent of the pot. This is a 20 percent cost compared with a 16 percent chance of winning the pot. The risk is greater than the payoff, so Louise folds. Mike and Vince, however, have more outs, so pot odds dictate that they stick around.

Then comes the turn: the Jack of Clubs. As shown in Table 4-5, after the turn, with only one card left to go, Mike's pot odds are no longer better than his chances of drawing a winning card, and he folds. Though Vince starts out with a potentially better hand than Mike, he too eventually folds when the pot odds indicate he should.

Table 4-5. Analysis after the turn

Player	Thelma	Louise	Mike	Vince
Needed cards			Same as before	Same as before
Chance of getting card			18 percent	20 percent
Current pot	350		450	450
Cost to call as percentage of pot			22 percent	22 percent
Action	Bet 100		Fold	Fold

Let's assume that the players are using only pot odds to make their decisions, ignoring for the sake of illustration that they are probably trying to get a read on the other players (e.g., who could bluff, raise, and so on). By the way, players are calculating the chance that they will get a card to improve their hand using the rule of four and the rule of 2 + 2 [Hack #36].

Why It Works

Imagine a game that costs a dollar to play. Pretend the rules are such that half the time you will win and get paid three dollars. The other half of the time you would lose one dollar and gain two dollars. Over time, if you kept playing this crazy game, you would make a whole lot of money.

It is the same sort of thinking that governs the use of pot odds in poker. With a 36 percent chance of making a flush, a perfectly fair bet would be to wager 36 percent of the pot. You would get your flush 36 percent of the time and break even over the long run. If you could play a game in which you could pay less than 36 percent of the pot and still win 36 percent of the time in the long run, you should play that crazy game, right? Well, every time you find yourself in a situation in which the pot odds are better than the proportion of the pot you have to wager, you have an opportunity to play just such a crazy game. Trust the statistics. Play the crazy game.

Where Else It Works

Experienced players not only make use of pot odds to make decisions about folding their hands, but they even make use of a slightly more sophisticated concept known as *implied pot odds*. Implied pot odds are based not on the proportion of the current pot that a player must call, but on the proportion of the pot total when the betting is completed for that betting round.

If players have yet to act, a player who is undecided about whether to stay in based on pot odds might expect other players to call down the line. This increases the amount of the final pot, increases the amount the player would win if she hit one of her wish cards, and increases the actual pot odds when all the wagering is done.

The phrase "implied pot odds" is also sometimes used to refer to the relative cost of betting compared to the final, total pot after all rounds of betting have been completed. I have also heard the term "pot odds" used to describe the idea that if you happen to "hit the nuts" (get a strong hand that's unlikely to be beaten) or close to it, then you are likely to win a pot much bigger than the typical pot. Some players spend a lot of energy and a lot of calls just hoping to hit one of these super hands and really clean up.

Implied pot odds works like this. In the scenario in Table 4-3, Mike might have called after Fourth Street (the fourth card revealed), anticipating that Vince would also call. This would have increased the final pot to 650, making Mike's contribution that round only 15 percent and justifying his call.

Interestingly, if Vince had been betting into a slightly larger pot that contained Mike's call, the pot odds for Vince's 100-chip call would then have dropped to 18 percent and Vince might have called. In fact, if Mike were a super genius-type player, he well could have called on the turn knowing that would change the pot odds for Vince and therefore encourage him to call. Real-life professional poker players—who are really, really good—really do think that way sometimes.

Where It Doesn't Work

Remember that pot odds are based on the assumption that you will be playing poker for an infinite amount of time. If you are in a no-limit tournament format, though, where you can't dig into your pockets, you might not be willing to risk all or most of your chips on your faith about what will happen in the long run.

The other problem with basing life and death decisions on pot odds is that you are treating a "really good hand" as if it were a guaranteed winner. Of course, it's not. The other players may have really good hands, too, that are better than yours.

Know When to Walk Away

#38   In Texas Hold 'Em, when you are "short-stacked," you have only a couple of choices: go all-in right now or go all-in very soon. As you might have guessed, knowing when to make your last stand is all about the odds.

I hear the TV poker commentators talking about how "easy" it is in Texas Hold 'Em tournaments to play when you are *short-stacked*. They mean it is easy because you don't have many options from which to choose.

The term "short-stacked" can be used in a couple of different ways. Sometimes, it is used to refer to whoever has the fewest chips at the table. Under this use of the term, even if you have thousands of chips and can afford to pay a hundred antes and big blinds, you are short-stacked if everyone else has more chips.

A better definition, which is more applicable to statistics-based decision making, is that you are short-stacked when you can only afford to pay the antes and blinds for a few more times around the table. Under this definition, there is mounting pressure to bet it all and hope to double or triple up and get back in the game. I prefer this use of the term because without pressure to play, being "short-stacked" is not a particularly meaningful situation.

It doesn't feel easy, though, does it, when you are short-stacked and have to go *all-in* (bet everything you have)? It feels very, very hard for two reasons:

- You are probably not going to win the tournament. You realize that you are down to very few chips and would have to double up several times to get back in the game. Realistically, you doubt that you have much of a chance. That's depressing, and any decision you make when you are sad is difficult.

- One mistake and you are out. There is little margin for error, and it is hard to pull the trigger in such a high-stakes situation.

Applying some basic statistical principles to the decision might help make you feel better. At least you'll have some nonemotional guidelines to follow. When you lose (and you still probably will; you're short-stacked, after all), now you can blame me, or the fates, and not yourself.

Recognizing a Short-Stacked Situation

In tournament settings, at some point you often will have so few chips that you will run out soon. Unless you bet and win soon, you will be *blinded out*—the cost of the mandatory bets will bleed you dry.

How few chips must you have to be short-stacked? Even if we define short-stacked as having some multiple of the *big blind* (the larger of two forced bets that you must make on a rotating basis), how *many* of those big blinds you need is a matter of style, and there is no single correct number. Here are some different perspectives on how many chips you must have in front of you to consider yourself short-stacked.

Twelve times the big blind or less. Though you could play quite a while longer without running out of chips, you will want to bet on any decent hand. You hope to win some blinds here. The more blinds you win, the longer you can wait for killer hands. If you are raised, at least consider responding with an all-in.

Players who start to think of themselves as short-stacked in this position wish to go all-in now on a good hand, rather than being forced to go all-in on a mediocre hand later on. Another advantage of starting to take risks is that an announcement of "all-in" will still pull some weight here. You will have enough chips to make someone think twice before they call you. Later on, your miserable little stack won't be enough to push anyone around.

> Choose your opponent wisely, if you can, when you go all-in and want a fold in response. Your raise of all-in against another small stack will be much more powerful than the same tactic against a monster stack. By the same token, if you want a call, don't hesitate to go all-in against players with tons of chips. They will be more than happy to double you up.

Eight times the big blind or less. In any position, whether you are on the button, in the big blind, or the first to bet, consider announcing all-in with any top-10 hand. You still have enough chips here to scare off some players, especially those with similarly sized stacks.

You are starting to get low enough, though, that you really want to be called. If you can play some low pairs cheaply, try it, but bail out if you don't get three of a kind in the flop. You need to keep as many big blinds as you can to coast on until you get that *all-in* opportunity.

Here are the 10 hands that are the most likely to double you up:

- A pair of Aces, Kings, Queens, Jacks, or 10s

- Ace-King, Ace-Queen, Ace-Jack, or King-Queen of the same suit
- Ace-King of different suits

Four times the big blind or less. At this point, you need to go all-in, even on hands that have a more than 50 percent chance of losing. Purposefully making a bad wager seems counterintuitive, but you are fighting against the ever-shrinking base amount you hope to double up. If you wait and wait until you have close to a sure thing, whatever stack remains will have to be doubled a few extra times to get you back.

A form of pot odds [Hack #37] kicks in at this point. If you pass up a 25 percent chance of winning while waiting for a 50 percent chance, you might be able to win only half as much when (and if) you ever get to play the better hand. Definitely go all-in on any pair, an Ace and anything else, any face card and a good kicker, or suited connectors.

> A good rule of thumb when you're very, very short-stacked (i.e., your total chips are fewer than four times the big blind) is to bet it all as soon as you get a hand that adds up to 18 or better. Kings count as 13, Queens 12, Jacks 11, and the rest are their face value. Aces count as 14, but you are already going all-in with an Ace-anything, so that doesn't matter. Eighteen-point hands include 10-8, Jack-7, Queen-6, and King-5.

Statistical Decision Making

The statistical question that determines when you should make your move—whether it is announcing all-in or, at least, making a decision to be *pot committed* (so many chips in the pot that you will go all-in if pushed)— is, "Am I likely to get a better hand before I run out of chips?"

I'm going to group 50 decent, playable starting Texas Hold 'Em poker hands, hands that give you a chance to win against a small number of opponents. I'll be using three groupings, shown in Tables 4-6, 4-7, and 4-8. While different poker experts might quibble a bit about whether a given hand is good or just okay, most would agree that these hands are all at least playable and should be considered when short-stacked.

> By the way, these hands are not necessarily in order of quality within each grouping.

Table 4-6. Ten great starting hands

Pairs	Same suit	Different suits
Ace-Ace	Ace-King	Ace-King
King-King	Ace-Queen	
Queen-Queen	Ace-Jack	
Jack-Jack	King-Queen	
10-10		

Table 4-7. Fifteen good starting hands

Pairs	Same suit	Different suits
9-9	Ace-Ten	Ace-Queen
8-8	King-Jack	Ace-Jack
7-7	King-10	King-Queen
	Queen-Jack	
	Queen-10	
	Jack-10	
	Jack-9	
	10-9	
	9-8	

Table 4-8. Twenty-five okay starting hands

Pairs	Same suit	Different suits
6-6	Ace-9	Ace-10
5-5	Ace-8	King-Jack
	Ace-7	Queen-Jack
	Ace-6	King-10
	Ace-5	Queen-10
	Ace-4	Jack-10
	Ace-3	
	Ace-2	
	King-9	
	Queen-9	
	10-8	
	9-7	
	8-7	
	8-6	
	7-6	
	6-5	
	5-4	

When you are short-stacked and the blinds and antes are coming due, you know you have a certain number of hands left before you have to make a move. Table 4-9 shows the probability that you will be dealt a great, good, or okay hand over the next certain number of deals.

Table 4-9. Chance of getting a playable hand

Hand quality	Next hand	In 5 deals	In 10 deals	In 15 deals	In 20 deals
Great	4 percent	20 percent	36 percent	49 percent	59 percent
Good	7 percent	29 percent	50 percent	65 percent	75 percent
Okay	11 percent	46 percent	70 percent	84 percent	91 percent
Okay or better	22 percent	72 percent	92 percent	98 percent	99 percent

I calculated the probabilities for Table 4-9 by first figuring the probabilities for any *specific* pair (you are just as likely to get a pair of Aces as a pair of 2s): .0045. I then figured the probabilities for getting any two *specific* different cards that are the same suit (.003), and the chances of getting any two *specific* different cards that are not the same suit (.009). Then, for each category—great, good, or okay—I multiplied the appropriate probability by the number of pairs, unpaired suited hands, and so on, in that category. I then calculated the chance of one of these hands *not* hitting across the given number of opportunities and subtracted that value from 1 to get the values for each cell in the table.

Here is how to use Table 4-9. Imagine you are short-stacked and have just been dealt a *good* hand. If you think you really have to go all-in sometime during the next five hands, there is only a 20 percent chance that you will be dealt a better hand. You should probably stake everything on this good hand.

If you can hang on for 20 more deals, there is a greater than 50 percent chance that you will get a gangbuster hand, so if you want to be conservative, you can lay these cards down for now. More commonly, short-stacked players consider going all-in with a hand that is not even a top-50 hand—something like King-8 unsuited, for example. Using the probabilities in Table 4-9, you might safely lay it down and hope for a better hand in the next five hands. There is a 72 percent chance you will get it.

Finally, imagine that you have just a few hands left because the blinds are shrinking your stack down to nothing. You look down and see a decent hand, an *okay* hand, such as 8-7 in the same suit. Table 4-9 allows you to answer the big question: is it likely that your very next hand will be better than this one? There is about an 11 percent chance of getting a good or great hand next. So, no, it is unlikely you will improve. Stake your future on this hand.

Getting Your Mind Right

We talked earlier about why it is so emotionally difficult to play when short-stacked. Here are some psychological tips to fight the pain of being caught between a rock and a hard place:

Be realistic

In blackjack, when a player hits her 16 against the dealer's 7, she knows she is likely to bust. She does it anyway because the dealer is likely to have a 10-card down, and it gives her the best chances in an almost no-win situation. She takes pleasure in knowing she did all she could to give herself the best chances of surviving. The same thinking applies here: take pleasure in knowing you gave yourself the best chances to come back and win.

Enjoy the all-in experience

There is nothing more exciting than having it all on the line. Because you had no real choice about going all-in, just relax and enjoy it the best you can. No player will chide you about doing "such a stupid thing," because you just did the smartest thing you could.

Take control

To avoid feeling forced to do something you don't want to do, start your comeback attempt before you have to. Play to avoid the short-stacked situation by starting to make your moves when you still have 10 to 12 times the big blind in chips. You have a lot more choices at this point than you will have later on, and so you can play with more sub-tlety, basing your bet on position, opponents, tells, and so on. The smaller your stack gets, the less power you have to control your own destiny.

HACK #39 Lose Slowly at Roulette

Roulette has so many pretty colors and shiny objects that kittens love it. Plus, you'll look pretty cool playing it. But in the long run, you'll lose money, and with your cat allergy and all....

Like most games in a casino, roulette is a game of pure chance. No one has any skill when it comes to predicting which of the 37 (European-style) or 38 (U.S.-style) partitioned sections the tiny ball will end up in. The best a player can do is know the odds, manage his money, and assume going in that he will lose.

Of course, he might get lucky and win some money, which would be dandy, but the Law of Big Numbers [Hack #2] must be obeyed. In the long run, he is most likely to have less money than if he had never played at all. In fact, if he plays an infinite amount of time, he is guaranteed to lose money. (Most roulette players play for a period of time somewhat less than infinity, of course.) To extend your amount of playing time, there is important statistical information you should know about this game with the spinning wheel, the orbiting ball, and the black and red layout.

Basic Wagers

Figure 4-1 shows the betting layout of a typical roulette game. This is an American-style layout, which means there are two green numbers, 0 and 00, which do not pay off any bets on red and black or odd and even. European-style roulette wheels have only one green number, 0, which cuts in half the house advantage compared to U.S. casinos.

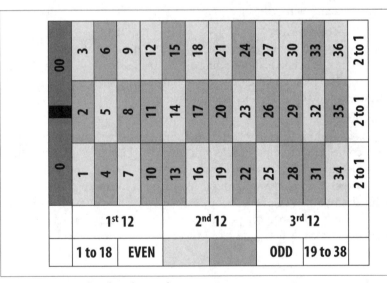

Figure 4-1. Typical roulette betting layout

Players can bet in a large variety of ways, which is one reason roulette is so popular in casinos. For example, a player could place one chip over a single number, touching two numbers, on a color, adjacent to a column of 12 numbers, and so on. Like any other probability question, the chance of randomly getting the desired outcome is a function of the number of desired outcomes (winning) divided by the total number of outcomes.

There are 38 spaces on the wheel and, because all 38 possible outcomes are equally likely, the calculations are fairly straightforward. Table 4-10 shows the types of bets players can make, the information necessary to calculate the odds of winning for a single spin of the wheel and a one-dollar bet, the actual amounts the casino pays out, and the house advantage.

Table 4-10. Statistics of roulette for each $1 bet

Type of wager	Number of winning outcomes	Number of losing outcomes	Odds	Casino pays	House edge equation	House edge
Single number	1	37	37 to 1	$35	$\frac{37-35}{38/1}$	5.26 percent
Two numbers	2	36	36 to 2 or 18 to 1	$17	$\frac{18-17}{38/2}$	5.26 percent
Single color	18	20	20 to 18 or 1.11 to 1	$1	$\frac{1.11-1}{38/18}$	5.26 percent
Even or odd	18	20	20 to 18 or 1.11 to 1	$1	$\frac{1.11-1}{38/18}$	5.26 percent
Twelve numbers	12	26	26 to 12 or 2.17 to 1	$2	$\frac{2.17-2}{38/12}$	5.26 percent

The house advantage is figured by first determining what the casino should pay back for each dollar bet if there were no advantage to the casino. The fair payback would be to give the winner an amount of money equal to the risk taken. The amount of risk taken is, essentially, the number of possible losing outcomes. This actual amount paid to the winner is then subtracted from the amount that should be paid if there were no house advantage. These "extra" dollars that the house keeps is divided by the proportion of total outcomes to winning outcomes. If there are no extra dollars, the game is evenly matched between player and casino and the house edge is 0 percent.

If you study the statistics of roulette in Table 4-10, a couple of conclusions are apparent. First, the casino makes its profit by pretending that there are only 36 numbers on a roulette wheel (i.e., only 36 possible outcomes) and pays out using that pretend distribution.

Second, regardless of the type of wager that is made at a roulette wheel, the house edge is a constant 5.26 percent. This is true except for one obscure wager, which is allowed at most casinos. Players are often allowed to bet on the two zeros and their adjacent numbers, 1, 2 and 3, for a total of five numbers. This is done by placing a chip to the side, touching both the 0 and the 1. I'd tell you more about checking with the person who spins the wheel to make sure they take this wager, and so on, except that this is the worst bet at the roulette table and no statistician would advise it. Casinos who allow this bet pay out as if it were a bet on six numbers. So, the casino's usual edge of 5.26 percent is even larger here: 7.89 percent, as shown in Table 4-11.

Table 4-11. Statistics for betting on five numbers in roulette (an inadvisable wager)

Type of wager	Number of winning outcomes	Number of losing outcomes	Odds	Casino pays	House edge equation	House edge
Five numbers	5	33	33 to 5 or 6.6 to 1	$6	$\dfrac{6.6 - 6}{38/5}$	7.89 percent

Why It Works

Roulette's popularity is based partly on the fact that so many different types of wagers are possible. A gambler with a lot of chips can spread them out all over the table, with a wide variety of different bets on different numbers and combinations of numbers. As long as she avoids the worst bet at the table (five numbers), she can rest assured that the advantage to the house will be the same honest 5.26 percent for each of her bets. It is one less thing for the gambler to worry about.

The fact that there is such a large variety of bets that can be placed on a single layout is no lucky happenstance, though. The decision to use 36 numbers was a wise one, and no doubt it was made all those years ago because of the large number of factors that go into 36. Thirty-six can be evenly divided by 1, of course, but also by 2, 3, 4, 6, 9, 12, and 18, making so many simple bets possible.

Play in the Black in Blackjack

HACK
#40

Perhaps the most potentially profitable application of statistics hacking is at the blackjack table.

In blackjack, the object of the game is to get a hand of cards that is closer to totaling 21 points (without going over) than the dealer's cards. It's a simple game, really. You start with two cards and can ask for as many more as you would like. Cards are worth their face value, with the exception of face cards, which are worth 10 points, and Aces, which can be either 1 or 11.

You lose if you go over 21 or if the dealer is closer than you (without going over). The bets are even money, with the exception of getting a *blackjack*: two cards that add up to 21. Typically, you get paid 3-to-2 for hitting a blackjack. The dealer has an advantage in that she doesn't have to act until after you. If you *bust* (go over 21), she wins automatically.

Statisticians can play this game wisely by using two sources of information: the dealer's face-up card and the knowledge of cards previously dealt. Basic strategies based on probability will let smart players play almost even against the house without having to pay much attention or learn complicated systems. Methods of taking into account previously dealt cards are collectively called *counting cards,* and using these methods allows players to have a statistical advantage over the house.

U.S. courts have ruled that card counting is legal in casinos, though casinos wish you would not do it. If they decide that you are counting cards, they might ask you to leave that game and play some other game, or they might ban you from the casino entirely. It is their right to do this.

Basic Strategy

First things first. Table 4-12 presents the proper basic blackjack play, depending on the two-card hand you are dealt and the dealer's up card. Most casinos allow you to *split* your hand (take a pair and split it into two different hands) and *double down* (double your bet in exchange for the limitation of receiving just one more card). Whether you should stay, take a card, split, or double down depends on the likelihood that you will improve or hurt your hand and the likelihood that the dealer will bust.

Table 4-12. Basic blackjack strategy against dealer's up card

Your hand	Hit	Stay	Double down	Split
5–8	Always			
9	2, 7–A		3–6	
10–11	10 or A		2–9	
12	2, 3, 7–A	4–6		
13–16	7–A	2–6		
17–20		Always		
2, 2	8–A			2–7
3, 3	2, 8–A			3–7
4, 4	2-5, 7–A			6
5, 5	10 or A		2–9	
6, 6	7–A			2–6
7, 7	8–A			2–7
8, 8				Always
9, 9	2–6, 8, 9			7, 10, A
10, 10		Always		

Table 4-12. Basic blackjack strategy against dealer's up card (continued)

Your hand	Hit	Stay	Double down	Split
A, A				Always
A, 2	2–5, 7–A		6	
A, 3 or A, 4	2–4, 7–A		5 or 6	
A, 5	2 or 3, 7–A		4–6	
A, 6	2, 7–A		3–6	
A, 7	9–A	2, 7–A	3–6	
A, 8 or 9 or 10		Always		

In Table 4-12, "Your hand" is the two-card hand you have been dealt. For example, "5–8" means your two cards total to a 5, 6, 7, or 8. "A" means Ace. A blank table cell indicates that you should never choose this option, or, in the case of splitting, that it is not even allowed.

The remaining four columns present the typical options and what the dealer's card should be for you to choose each option. As you can see, for most hands there are only a couple of options that make any statistical sense to choose. The table shows the best move, but not all casinos allow you to double-down on just any hand. Most, however, allow you to split any matching pair of cards.

Why It Works

The probabilities associated with the decisions in Table 4-12 are generated from a few central rules:

- The dealer is required to hit until she makes it to 17 or higher.
- If you bust, you lose.
- If the dealer busts and you have not, you win.

The primary strategy, then, is to not risk busting if the dealer is likely to bust. Conversely, if the dealer is likely to have a nice hand, such as 20, you should try to improve your hand. The option that gives you the greatest chance of winning is the one indicated in Table 4-12.

The recommendations presented here are based on a variety of commonly available tables that have calculated the probabilities of certain outcomes occurring. The statistics have either been generated mathematically or have been produced by simulating millions of blackjack hands with a computer.

Here's a simple example of how the probabilities battle each other when the dealer has a 6 showing. The dealer could have a 10 down. This is actually the most likely possibility, since face cards count as 10. If there is a 10 down, great, because if the dealer starts with a 16, she will bust about 62 percent of the time (as will you if you hit a 16).

Since eight different cards will bust a 16 (6, 7, 8, 9, 10, Jack, Queen, and King), the calculations look like this:

$$8/13 = .616$$

Of course, even though the *single* best guess is that the dealer has a 10 down, there is actually a better chance that the dealer does *not* have a 10 down. All the other possibilities (9/13) add up to more than the chances of a 10 (4/13).

Any card other than an Ace will result in the dealer hitting. And the chances of that next card breaking the dealer depends on the probabilities associated with the starting hand the dealer actually has. Put it all together and the dealer does not have a 62 percent chance of busting with a 6 showing. The actual frequency with which a dealer busts with a 6 showing is closer to 42 percent, meaning there is a 58 percent chance she will not bust.

Now, imagine that you have a 16 against the dealer's down card of 6. Your chance of busting when you take a card is 62 percent. Compare that 62 percent chance of an immediate loss to the dealer's chance of beating a 16, which is 58 percent. Because there is a greater chance that you will lose by hitting than that you will lose by not hitting (62 is greater than 58), you should stay against the 6, as Table 4-12 indicates.

All the branching possibilities for all the different permutations of starting hands versus dealers' up cards result in the recommendations in Table 4-12.

Sucker Bet

Many casinos offer a chance for you to buy *insurance* if the dealer's up card is an Ace. Insurance means that you wager up to half your original bet, and if the dealer has a blackjack (a 10 or face card as the down card), you win that side bet but lose your original wager (unless you, too, have a blackjack, in which case it's a tie and you get your wager back).

The chances of the dealer having a 10 underneath are 4/13, or 31 percent. You will lose your insurance money much more often then you will win it. Unless you are counting cards, never take insurance. Yes, even if you have a blackjack.

Simple Card-Counting Methods

The basic strategies described earlier in this hack assume that you have no idea what cards still remain in the deck. They assume that the original distribution of cards still remains for a single deck, or six decks, or whatever number of decks is used in a particular game. The moment any cards have been dealt, however, the actual odds change, and, if you know the new odds, you might choose different options for how you play your hand.

Elaborate and very sound (statistically speaking) methods exist for keeping track of cards previously dealt. If you are serious about learning these techniques and dedicating yourself to the life of a card counter, more power to you. I don't have the space to offer a complete, comprehensive system here, though. For the rest of us, who would like to dabble a bit in ways to increase our odds, there are a few counting procedures that will improve your chances without you having to work particularly hard or memorize many charts and tables.

The basic method for improving your chances against the casino is to increase your wager when there is a better chance of winning. The wager must be placed before you get to see your cards, so you need to know ahead of time when your odds have improved. The following three methods for knowing when to increase your bet are presented in order of complexity.

Counting Aces. You get even money for all wins, except when you are dealt a blackjack. You get a 3-to-2 payout (e.g., $15 for every $10 bet) when a blackjack comes your way. Consequently, when there is a better-than-average chance of getting a blackjack, you would like to have a larger-than-average wager on the line.

The chances of getting a blackjack, all things being equal, is calculated by summing two probabilities:

Getting a 10-card first and then an Ace
$4/13 \times 4/51 = .0241$

Getting an Ace first and then a 10-card
$1/13 \times 16/51 = .0241$

Add the two probabilities together, and you get a .0482 (about 5 percent) probability of being dealt a natural 21.

Obviously, you can't get a blackjack unless there are Aces in the deck. When they are gone, you have no chance for a blackjack. When there are relatively few of them, you have less than the normal chance of a blackjack. With one deck, a previously dealt Ace lowers your chances of hitting a blackjack to .0362 (about 3.6 percent). Dealing a quarter of the deck with no Aces showing up increases your chances of a blackjack to about 6.5 percent.

 Quick tip for the budding card counter: don't move your lips.

Counting Aces and 10s. Of course, just as you need an Ace to hit a blackjack, you also need a 10-card, such as a 10, Jack, Queen, or King. While you are counting Aces, you could also count how many 10-cards go by.

There is a total of 20 Aces and 10-cards, which is about 38 percent of the total number of cards. When half the deck is gone, half of those cards should have been shown. If fewer than 10 of these key cards have been dealt, your chances of a blackjack have increased. With all 20 still remaining halfway through a deck, your chances of seeing a blackjack in front of you skyrockets to 19.7 percent.

Going by the point system. Because you want proportionately more high cards and proportionately fewer low cards when you play, a simple point system can be used to keep a running "count" of the deck or decks. This requires more mental energy and concentration than simply counting Aces or counting Aces, 10s, and face cards, but it provides a more precise index of when a deck is loaded with those magic high cards.

Table 4-13 shows the point value of each card in a deck under this point system.

Table 4-13. Simple card-counting point system

Card	Point value
10, Jack, Queen, King, Ace	−1
7, 8, 9	0
2, 3, 4, 5, 6	+1

A new deck begins with a count of 0, because there are an equal number of −1 cards and +1 cards dealt in the deck. Seeing high cards is bad, because your chances of blackjacks have dropped, so you lose a point in your count. Spotting low cards is good, because there are now proportionately more high cards in the deck, so you gain a point there.

You can learn to count more quickly and easily by learning to rapidly recognize the total point value of common pairs of cards. Pairs of cards with both a high card and a low card cancel each other out, so you can quickly process and ignore those sorts of hands. Pairs that are low-low are worth big points (2), and pairs that are high-high are trouble, meaning you can subtract 2 points for each of these disappointing combinations.

You will only occasionally see runs of cards that dramatically change the count in the good direction. The count seldom gets very far from 0. For example, with a single new deck, the first six cards will be low less than 1 percent of the time, and the first ten cards will be low about 1/1000 of 1 percent of the time.

The count doesn't have to be very high, though, to improve your odds enough to surpass the almost even chance you have just following basic strategy. With one deck, counts of +2 are large enough to meaningfully improve your chances of winning. With more than one deck, divide your count by the number of decks—this is a good estimation of the true count.

Sometimes you will see very high counts, even with single decks. When you see that sort of string of luck, don't hesitate to raise your bet. If you get very comfortable with the point system and have read more about such systems, you can even begin to change the decisions you make when hitting or standing or splitting or doubling down.

Even if you just use these simple systems, you will improve your chances of winning money at the blackjack tables. Remember, though, that even with these sorts of systems, there are other pitfalls awaiting you in the casino, so be sure to always follow other good gambling advice [Hack #35] as well.

HACK #41 Play Smart When You Play the Lottery

Your odds of winning a big prize in a giant lottery are really, really small, no matter how you slice it. You do have some control over your fate, however. Here are some ways to give yourself an advantage (albeit slight) over all the other lotto players who haven't bought this book.

In October of 2005, the biggest Powerball lottery winner ever was crowned and awarded $340 million. It wasn't me. I don't play the lottery because, as a statistician, I know that playing only slightly increases my chances of winning. It's not worth it to me.

Of course, if I don't play, I can't win. Buying a lottery ticket isn't necessarily a bad bet, and if you are going to play, there are a few things you can do to increase the amount of money you will win (probably) and increase your chances of winning (possibly). Whoever bought the winning $340 million ticket in Jacksonville, Oregon, that October day likely followed a few of these winning strategies, and you should too.

Because Powerball is a lottery game played in most U.S. states, we will use it as our example. This hack will work for any large lottery, though.

Powerball Odds

Powerball, like most lotteries, asks players to choose a set of numbers. Random numbers are then drawn, and if you match some or all of the numbers, you win money! To win the biggest prizes, you have to match lots of numbers. Because so many people play Powerball, many tickets are sold, and the prize money can get huge.

Of course, correctly picking all the winning numbers is hard to do, but it's what you need to do to win the jackpot. In Powerball, you choose five numbers and then a sixth number: the red *powerball*. The regular white numbers can range from 1 to 55, and the powerball can range from 1 to 42. Table 4-14 shows the different combinations of matches that result in a prize, the amount of the prize, and the odds and probability of winning the prize.

Table 4-14. Powerball payoffs

Match	Cash	Odds	Percentage
Powerball only	$3	1 in 69	1.4 percent
1 white ball and the powerball	$4	1 in 127	0.8 percent
3 white balls	$7	1 in 291	0.3 percent
2 white balls and the powerball	$7	1 in 745	0.1 percent
3 white balls and the powerball	$100	1 in 11,927	0.008 percent
4 white balls	$100	1 in 14,254	0.007 percent
4 white balls and the powerball	$10,000	1 in 584,432	0.0002 percent
5 white balls	$200,000	1 in 3,563,609	0.00003 percent
5 white balls and the powerball	Grand prize	1 in 146,107,962	0.0000006 percent

Powerball Payoff

Armed with all the wisdom you likely now have as a statistician (unless this is the first hack you turned to in this book), you might have already made a few interesting observations about this payoff schedule.

Easiest prize. The easiest prize to win is the *powerball only* match, and even then there are slim chances of winning. If you match the powerball (and no other numbers), you win $3. The chances of winning this prize are about 1 in 69.

This is not a good bet by any reasonable standard. It costs a dollar to buy a ticket, to play one time, and the expected payout schedule is $3 for every 69 tickets you buy. So, on average, after 69 plays you will have won $3 and spent $69.

Actually, your payoff will be a little better than that. The odds shown in Table 4-14 are for making a specific match and not doing any better than that. Some proportion of the time when you match the powerball, you will also match a white ball and your payoff will be $4, not $3. Choosing five white ball numbers and matching at least 1 will happen 39 percent of the time.

So, after having matched the powerball, you have a little better than a third chance of hitting at least one white ball as well. Even so, your expected payoff is about $3.39 for every $69 you throw down that rat hole (I mean, spend on the lottery), which is still not a good bet.

Powerball only. The odds for the *powerball only* match don't seem quite right. I said there were 42 different numbers to choose from for the powerball, so shouldn't there be 1 out of 42 chances to match it, not 1 in 69?

Yes, but remember this shows the chances of hitting that prize only and not doing better (by matching some other balls). Your odds of winning something, anything, if you combine all the winning permutations together are 1 in 37, about 3 percent. Still not a good bet.

Grand prize. The odds for the grand prize don't seem quite right either. (Okay, okay, I don't really expect you to have "noticed" that. I didn't either until I did a few calculations.)

If there are 5 draws from the numbers of 1 to 55 (the white balls) and 1 draw from the numbers 1 to 42 (the red ball), then a quick calculation would estimate the number of possibilities as:

$$(55)(55)(55)(55)(55)(42) = 21,137,943,750$$

In other words, the odds are 1 out of 21,137,943,750. Or, if you were thinking a little more clearly, realizing that the number of balls gets smaller as they are drawn, you might speedily calculate the number of possible outcomes as:

$$(55)(54)(53)(52)(51)(42) = 17,532,955,440$$

But the odds as shown are somewhat better than 1 out of 1.7 billion. The first time I calculated the odds, I didn't keep in mind that the order doesn't matter, so any of the remaining chosen numbers could come up at any time. Hence, here's the correct series of calculations:

$$(5/55)(4/54)(3/53)(2/52)(1/51)(1/42) = 1/146,107,962$$

Winning Powerball

OK, Mr. Big Shot Stats Guy (you are probably thinking), you're going to tell us that we should never play the lottery because, statistically, the odds will never be in our favor. Actually, using the criteria of a fair payout, there is one time to play and to buy as many tickets as you can afford.

In the case of Powerball, you should play anytime the grand prize increases to past $146,107,962 (or double that amount if you want the lump sum payout). As soon as it hits $146,107, 963, buy, buy, buy! Because the chances of matching five white balls and the one red ball are exactly one out of that big number, from a statistical perspective, it is a good bet anytime your payout is bigger than that big number.

For Powerball and its number of balls and their range of values, 146,107,962 is the magic number. The idea that your chances of winning haven't changed but the payoff amount has increased to a level where playing is worthwhile is similar to the concept of pot odds in poker [Hack #37].

You can calculate the "magic number" for any lottery. Once the payoff in that lottery gets above your magic number, you can justify a ticket purchase. Use the "correct series" of calculations in our example for Powerball as your mathematical guide. Ask yourself how many numbers you must match and what the range of possible numbers is. Remember to lower the number you divide by one each time you "draw" out another ball or number, unless numbers can repeat. If numbers can repeat, then the denominator stays the same in your series of multiplications.

One important hint about deciding when to buy lottery tickets has to do with determining the *actual* magic number, the prize amount, which triggers your buying spree. The amount that is advertised as the jackpot is not, in fact, the jackpot. The advertised "jackpot" is the amount that the winner would get over a period of years in a regular series of smaller portions of that amount. The *real* jackpot—the amount you should identify as the payout in the gambling and statistical sense—is the amount that you would get if you chose the *one lump sum* option. The one lump sum is typically a little less than half of the advertised jackpot amount.

So, if you have determined that your lottery has grown a jackpot amount that says it is now statistically a good time to play, how many tickets should you buy? Why not buy one of each? Why not spend $146,107,962 and buy every possible combination? You are guaranteed to win. If the jackpot is greater than that amount, then you'll make money, guaranteed, right? Well, actually not. Otherwise, I'd be rich and I would never share this hack with you. Why wouldn't you be guaranteed to win? The probably is that you might be forced to...wait for it...split the prize! Argh! See the next section...

Don't Split the Prize

If you do win the lottery, you'd like to be the only winner, so in addition to deciding when to play, there are a variety of strategies that increase the likelihood that you'll be the only one who picked your winning number.

First off, I'm working under the assumption that the winning number is randomly chosen. I tend not to be a conspiracy theorist, nor do I believe that God has the time or inclination to affect the drawing of winning lottery numbers, so I'm going to not list any strategy that would work only if there were not randomness in the drawing of lottery numbers. Here are some more reasonable tips to consider when picking your lottery numbers:

Let the computer pick
> Let the computer do the picking, or, at least, choose random numbers yourself. Random numbers are less likely to have meaning for any other player, so they are less likely to have chosen them on their own tickets. The Powerball people report that 70 percent of all winning tickets are chosen randomly by the in-store computer. (They also point out, in a bit of "We told you that results are random" whimsy that 70 percent of *all* tickets purchased had numbers generated by the computer.)

Don't pick dates
> Do not pick numbers that could be dates. If possible, avoid numbers lower than 32. Many players always play important dates, such as birthdays and anniversaries, prison release dates, and so on. If your winning number could be someone's lucky date, that increases the chance that you will have to split your winnings.

Stay away from well-known numbers
> Do not pick numbers that are well known. In the big October 2005 Powerball results, hundreds of players chose numbers that matched the lottery ticket numbers that play a large role in the popular fictional TV show *Lost*. None of these folks won the big prize, but if they had, they would have had to divide the millions into hundreds of slices.

There is also a family of purely philosophical tips that have to do with abstract theories of cause and effect and the nature of reality. For example, some philosophers would say to pick last week's winning numbers. Because, while you might not know for sure what is real and what can and cannot happen in this world, you do know that, at least, it is possible for last week's numbers to be this week's winning numbers. It happened before; it can happen again.

Though your odds of winning a giant lottery prize are slim, you can follow some statistical principles and do a few things to actually control your own destiny. (The word for destiny in Italian, by the way, is *lotto*.) Oh, and one more thing: buy your ticket on the *day* of the drawing. If too much time passes between your purchase and the announcement of the winning numbers, you have a greater likelihood of being hit by lightning, drowning in the bathtub, or being struck by a minivan than you do of winning the jackpot. Timing is everything, and I'd hate for you to miss out.

HACK #42 Play with Cards and Get Lucky

While it is true that Uncle Frank spends much of his time in taverns using dice to win silly bar bets and smiling real charming-like at the ladies, there is more to his life than that. For instance, sometimes he uses playing cards instead of dice.

People, especially card players, and especially poker players, feel pretty good about their level of understanding of the likelihood that different combination of cards will appear. Their experience has taught them the relative rarity of pairs, three-of-a-kind, flushes, and so on. Generalizing that intuitive knowledge to playing-card questions outside of game situations is difficult, however.

My stats-savvy uncle, Uncle Frank, knows this. Sometimes, Uncle Frank uses his knowledge of statistics for evil, not good, I am sorry to say, and he has perfected a group of bar bets using decks of playing cards, which he claims helped pay his way through graduate school. I'll share them with you only for the purpose of demonstrating certain basic statistical principles. I trust that you will use your newfound knowledge to entertain others, fight crime, or win inexpensive nonalcoholic beverages.

Getting a Li'l Flush

In poker, a flush is five cards, all of the same suit. For my Uncle Frank, though, there is seldom time to deal out complete poker hands before he is asked to leave whatever establishment he is in. Consequently, Uncle Frank often makes wagers based on what he calls *li'l flushes*.

The bet. A little flush (oops, sorry; I mean *li'l* flush) is any two cards of the same suit. Frank has a wager that he almost always wins that has to do with finding two cards of the same suit in your hand. Again, because of time constraints, his poker hands have only four cards, not five.

The wager is that you deal me four cards out of a random deck, and I will get at least two cards of the same suit. While this might not seem too likely, it is actually much less likely that there would be four cards of all *different* suits. I figure the chance of getting four different suits in a four-card hand is about 11 percent. So, the likelihood of getting a li'l flush is about 89 percent!

Why it works. There are a variety of ways to calculate playing-card hand probabilities. For this bar bet, I use a method that counts the number of possible winning hand combinations and compares it to the total number of hand combinations. This is the method used in "Play with Dice and Get Lucky" **[Hack #43]**.

To think about how often four cards would represent four different suits, with no two-card flushes amongst them, count the number of possible four-card hands. Imagine any first card (52 possibilities), imagine that card combined with any remaining second card (52×51), add a third card ($52 \times 51 \times 50$) and a fourth card ($52 \times 51 \times 50 \times 49$), and you'll get a total of 6,497,400 different four-card hands.

Next, imagine the first two cards of a four-card hand. These will match only .2352 of the time (12 cards of the same suit remain out of a 51-card deck). So, about one-and-a-half million four-card deals will find a flush in the first two cards. They won't match another .7648 of the time. This leaves 4,968,601 hands with two differently suited first two cards.

Of that number, how many will not receive a third card that does not suit up with either of the first two cards? There are 50 cards remaining, and 26 of those have suits that have not appeared yet. So, 26/50 (52 percent) of the time, the third card would not match either suit.

That leaves 2,583,673 hands that have three first cards that are all unsuited. Now, of that number, how many will now draw a fourth card that is the fourth unrepresented suit? There are 13 out of 49 cards remaining that represent that final fourth suit. 26.53 percent of the remaining hands will have that suit as the fourth card, which computes to 685,464 four-card combinations with four different suits. 685,464 divided by the total number of possible hands is .1055 (685,464/6497400).

There's your 11 percent chance of having four different suits in a four-card hand. Whew! By the way, some super-genius-type could get the same proportion by using just the relevant proportions, which we used along the way during our different counting steps, and not have to count at all:

$$.7648 \times .52 \times .2653 = .1055$$

Finding a Match with Two Decks of Cards

You have a deck of cards. I have a deck of cards. They are both shuffled (or, perhaps, souffléd, as my spell check suggested I meant to say). If we dealt them out one at a time and went through both decks one time, would they ever match? I mean, would they ever match *exactly*, with the exact same card—for example, us both turning up the Jack of Clubs at the same time?

The bet. Most people would say no, or at least that it would certainly happen occasionally, but not too frequently. Astoundingly, not only will you often find at least one match when you pass through a pair of decks, but it would be out of the ordinary *not* to. If you make this wager or conduct this experiment many times, you will get at least one match on most occasions. In fact, you will *not* find a match only 36.4 percent of the time!

Why it works. Here's how to think about this problem statistically. Because the decks are shuffled, one can assume that any two cards that are flipped up represent a random sample from a theoretical population of cards (the deck). The probability of a match for any given sample pair of cards can be calculated. Because you are sampling 52 times, the chance of getting a match *somewhere* in those attempts increases as you sample more and more pairs of cards. It is just like getting a 7 on a pair of dice: on any given roll, it is unlikely, but across many rolls, it becomes more likely.

To calculate the probability of hitting the outcome one wishes across a series of outcomes, the math is actually easier if one calculates the chances of *not* getting the outcome and multiplying across attempts. For any given card, there is a 1 out of 52 chance that the card in the other deck is an exact match. The chances of that not happening are 51 out of 52, or .9808.

You are trying to make a match more than once, though; you are trying 52 times. The probability of not getting a match across 52 attempts, then, is .9808 multiplied by itself 52 times. For you math types, that's $.9808^{52}$.

Wait a second and I'll calculate that in my head (.9808 times .9808 times .9808 and so on for 52 times is...about...0.3643). OK, so the chance that it *won't* happen is .3643. To get the chance that it *will* happen, we subtract that number from 1 and get .6357.

You'll find at least one match between two decks about two-thirds of the time! Remarkable. Go forth and win that free lemonade.

HACK #43   Play with Dice and Get Lucky

Here are some honest wagers using honest dice. Just because you aren't cheating, though, doesn't mean you won't win.

It is an unfortunate stereotype that statisticians are glasses-wearing introspective nerds who never have a beer with the gang. This is such an absurd belief, that just thinking about it last Saturday and Sunday at my weekly *Dungeons & Dragons* gathering, I laughed so hard that my monocle almost landed in my sherry.

The truth is that displaying knowledge of simple probabilities in a bar can be quite entertaining for the patrons and make you the life of the party. At least, that's what happens according to my Uncle Frank, who for years has used his stats skills to win free drinks and pickled eggs (or whatever those things are in that big jar that are always displayed in the bars I see on TV).

Here are a few ways to win a bet using any fair pair of dice.

Distribution of Dice Outcomes

First, let's get acquainted with the possibilities of two dice rolled once. You'll recall that most dice have six sides (my fantasy role-playing friends and I call these *six-sided dice*) and that the values range from 1 to 6 on each cube.

Calculating the possible outcomes is a matter of listing and counting them. Figure 4-2 shows all possible outcomes for rolling two dice.

1 way to roll **2**		2 ways to roll **3**		3 ways to roll **4**		4 ways to roll **5**		5 ways to roll **6**		6 ways to roll **7**		5 ways to roll **8**		4 ways to roll **9**		3 ways to roll **10**		2 ways to roll **11**		1 way to roll **12**	
1	1	1	2	1	3	1	4	1	5	1	6	2	6	3	6	4	6	5	6	6	6
		2	1	2	2	2	3	2	4	2	5	3	5	4	5	5	5	6	5		
				3	1	3	2	3	3	3	4	4	4	5	4	6	4				
						4	1	4	2	4	3	5	3	6	3						
								5	1	5	2	6	2								
										6	1										

Figure 4-2. Possible outcomes for two dice

This distribution results in the frequencies shown in Table 4-15.

Table 4-15. Frequency of outcomes for rolling two dice

Total roll	Chances	Frequency
2	1	2.8 percent
3	2	5.6 percent
4	3	8.3 percent
5	4	11.1 percent
6	5	13.9 percent
7	6	16.7 percent
8	5	13.9 percent
9	4	11.1 percent
10	3	8.3 percent
11	2	5.6 percent
12	1	2.8 percent
Total number of possible outcomes	36	100 percent

The game of craps, of course, is based entirely on these expected frequencies. Some interesting wagers might come to mind as you look at this frequency distribution. For example, while a 7 is the most common roll and many people know this, it is only slightly more likely to come up than a 6 or 8.

In fact, if you didn't have to be specific, you could wager that a 6 *or* an 8 will come up before a 7 does. Of all totals that could be showing when those dice are done rolling, more than one-fourth of the time (about 28 percent) the dice will total 6 or 8. This is substantially more likely than a 7, which comes up only one-sixth of the time.

Bar Bets with Dice

My Uncle Frank used to bet any dull-witted patron that he would roll a 5 or a 9 before the patron rolled a 7. Uncle Frank won 8 out of 14 times.

Sometimes, old Frankie would wager that on any one roll of a pair of dice, there would be a 6 or a 1 showing. Though, at first thought, there would seem to be at least a less than 50 percent chance of this happening, the truth is that a 1 or 6 will be showing about 56 percent of the time. This is the same probability for any two different numbers, by the way, so you could use an attractive stranger's birthday to pick the digits and maybe start a conversation, which could lead to marriage, children, or both.

If you are more honest than my Uncle Frank (and there is a 98 percent chance that you are), here are some even-money bets with dice. The outcomes in column A are equally as likely to occur as the outcomes in column B:

A	B
2 or 12	3
2, 3, or 4	7
5, 6, or 7	8, 9, 10, 11, or 12

The odds are even for either outcome.

Why It Works

For the bets presented in this hack, here are the calculations demonstrating the probability of winning:

Wager	Number of winning outcomes	Calculation	Resulting proportion
5 or 9 versus 7	8 versus 6	8/14	.571
1 or 6 showing	20	20/36	.556
2 or 12 versus 3	2 versus 2	2/4	.500
2, 3 or 4 versus 7	6 versus 6	6/12	.500
5, 6 or 7 versus 8 or higher	15 versus 15	15/30	.500

The "Wager" column presents the two competing outcomes (e.g., will a 5 or 9 come up before a 7?). The "Number of winning outcomes" column indicates number of different dice rolls that would result in either side of the wager (e.g., 8 chances of getting a 5 or 9 versus only 6 chances of getting a 7). The "Resulting proportion" column indicates your chances of winning.

You can win two different ways with these sorts of bets. If it is an even-money bet, you can wager less than your opponent and still make a profit in the long run. He won't know the odds are even. If chance favors you, though, consider offering your target a slightly better payoff, or pick the outcome that is likely to come up more often.

HACK #44   Sharpen Your Card-Sharping

In Texas Hold 'Em and other poker games, there are a few basic preliminary skills and a bit of basic knowledge about probability that will immediately push you from absolute beginner to the more comfortable level of knowing just enough to get into trouble as a card sharp.

The professional Texas Hold 'Em poker players who appear on television are different from you and me in just a couple of important ways. (Well, they likely differ from *you* in just a couple of important ways; they differ from me in so many important ways that even my computer brain can't count that high.) Here are two areas of poker playing that they have mastered:

- Knowing the rough probability of hitting the cards they want at different stages in a hand (in the flop, on the river, and so on)
- Quickly identifying the possible better hands that could be held by other players

This hack presents some tips and tools for moving from novice to semi-pro. These are some simple hunks of knowledge and quick rules of thumb for making decisions. Like the other poker hacks in this book, they provide strategy tips based purely on statistical probabilities, which assume a random distribution of cards in a standard 52-card deck.

Improving Your Hand

Half the time, you will get a pair or better in Texas Hold 'Em. I'll repeat that because it is so important in understanding the game. Half the time (a little under 52 percent actually), if you stay in long enough to see seven cards (your two cards plus all five community cards), you will have at least one pair. It might have been in your hand (a *pocket* or *wired* pair), it might be made up of one card in your hand and one from the community cards, or your pair might be entirely in the community cards for everyone to claim.

If for the majority of the time the average player will have a pair when dealt seven cards, then sticking around until the end with a low pair means you are—only statistically speaking, of course—likely to lose. In other words, there is a greater than 50 percent chance that the other player has *at least* a pair, and that pair will probably be 8s or higher (only six out of thirteen pairs are 7s or lower.)

Knowing how common pairs are explains why Aces are so highly valued. Much of the time, heads-up battles come down to a battle of pair versus pair. Another good proportion of the time, the Ace plays an important role as a kicker or tiebreaker. Aces are good to have, and it's all because of the odds.

Probabilities. Decisions about staying in or raising your bet in an attempt to lower the number of opponents you have to beat can be made more wisely if you know some of the common probabilities for some of the commonly hoped-for outcomes. Table 4-16 presents the probability of drawing a card that helps you at various stages in a hand. The probabilities are calculated based on how many cards are left in the deck, how many different cards will help you (your *outs*), and how many more cards will be drawn from the deck. For example, if you have an Ace-King and hope to pair up, there are six cards that can make that happen; in other words, you have six outs. If you have only an Ace high but hope to find another Ace, you have three outs. If you have a pocket pair and hope to find a powerful third in the community cards, you have just two outs.

Table 4-16. Probability of improving your hand

Cards left to be dealt	Six outs	Three outs	Two outs
5 (before the flop)	49 percent	28 percent	19 percent
2 (after the flop)	24 percent	12 percent	8 percent
1 (after the turn)	13 percent	7 percent	4 percent

The situations described here assume you have already been dealt two cards. After all, in most poker games, the bet before the flop is predetermined, so there are no decisions to be made. By the way, because you should probably back out of hands that did not amount to anything in the flop, you'll want to know your chances of improving in the flop itself. They are:

Remaining outs	Odds you'll hit a winning card in the flop
6	32 percent
3	17 percent
2	12 percent

Implications. Here are a few quick observations and implications to etch in your mind, based on the distribution described in Table 4-16.

Half the time, you will pair up. This is true for high cards, such as *Big Slick* (Ace-King) or low cards, such as 2-7. You can even pick from the two cards you have and pair that one up 28 percent of the time. Implication: when low on chips in tournament play, go all-in as soon as you get that Ace.

If you don't hit the third card, you need to turn a pair into a *set* (three of a kind) on the flop, and there is only an 8 percent chance you will hit it down the road. Implication: don't spend too much money waiting around for your low pair to turn into a gangbuster hand.

Your Ace-King or Ace-Queen that looked pretty good before the flop diminishes in potential as more cards are revealed without pairing up or getting straight draws. 87 out of 100 times, that great starting hand remains a measly high-card-only hand if you haven't hit before the river. Implication: stay in with the unfulfilled dream that is Ace-King only if you can do so cheaply.

Reading the Community Cards in a Flash

Here are some common-sense statements about your opponents' hands that must be true but aren't always said out loud:

If the community cards do not have...	Your opponent(s) cannot have...
A pair	Four of a kind
A pair	A full house
Three cards of the same suit	A flush
Three cards within a five-card range	A straight

You can make quicker decisions about what your opponents might have by learning these rules. Then, you can automatically rule out killer hands when the situation is such that they are impossible. You may not be worried about speed, but you can spend your time concentrating on more important decisions if you don't have to waste mental energy figuring these things out from scratch each time.

HACK #45   Amaze Your 23 Closest Friends

What are the chances of at least two people in a group sharing a birthday? Depending on the number of people present, surprisingly high. Impress your friends at parties (and perhaps win some money in a bar bet) using these simple rules of probability.

Some events that seem logically unlikely can actually turn out to be quite probable in some cases. One such example is determining the probability that at least two people in a group share a birthday. Many people are shocked to learn that as long as there are at least 23 people in the group, there is a better than 50 percent chance that at least 2 of them will have the same birthday! By using a few simple rules of probability, you can figure out the likelihood of this event occurring for groups of any size, and then amaze your friends when your predictions come true.

You could also use this result to make some cash in a bar bet (as long as there are at least 23 people there).

So, how do you figure out the probability of at least two people sharing a birthday? To solve this problem, you need to make a couple of assumptions about how birthdays are distributed in the population and know a few rules about how probabilities work.

Getting Started

To determine the chances of at least two people sharing a birthday, we have to make a couple of reasonable assumptions about how birthdays are distributed. First, let's assume that birthdays are uniformly distributed in the population. This means that approximately the same number of people are born on every single day of the year.

This is not necessarily perfectly true, but it's close enough for us to still trust our results. However, there is one birthday for which this is definitely not true: February 29, which occurs only every four years on Leap Year. The good news is that few enough people are born on February 29 that it is easy for us to just ignore it and still get accurate estimates.

Once we've made these two assumptions, we can solve the birthday problem with relative ease.

Applying the Law of Total Probability

In our problem, there are only two *mutually exclusive* possible outcomes:

- At least two people share a birthday.
- No one shares a birthday.

Since one of these two things *must* occur, the sum of the two probabilities will always be equal to one. Statisticians call this the *Law of Total Probability*, and it comes in handy for this problem.

The term *mutually exclusive* means that if one event occurs, the other one cannot occur, and vice versa.

A simple coin-flipping example can help picture how this works. With a fair coin, the probability of getting a *heads* is 0.5, just as the probability of getting a *tails* is 0.5 (which is another example of mutually exclusive events, because the coin can't come up heads *and* tails in the same flip!). Once you flip the coin, one of two things has to happen. It *must* land either heads up or tails up, so the probability of heads *or* tails occurring is 1 (0.5 + 0.5). Conversely, we can think of the probability of heads as one minus the probability of tails (1 − 0.5 = 0.5), and vice versa.

Sometimes, it's easier to determine the probability of an event *not* occurring and then use that information to determine the probability that it *will* occur. The probability of no one sharing a birthday is a bit easier to figure out, and it depends only on how many people are in the group.

Imagine that our group contains only two people. What's the probability that they share a birthday? Well, the probability that they *don't* share a birthday is easy to figure: person #1 has some birthday, and there are 364 other birthdays person #2 might have that would result in them not sharing a birthday. So, mathematically, it's 364 divided by 365 (the total number of possible birthdays), or 0.997.

Since the probability of the two people *not* sharing a birthday is 0.997 (a very high probability), the probability of them actually sharing a birthday is equal to $1 - 0.997$ (0.003, a very low probability). This means that only 3 out of every 1,000 randomly selected pairs of people will share a birthday. So far, this makes perfect logical sense. However, things change (and change quickly) once we start adding more people to our group!

Calculating the Probability of Independent Events

The other trick we need to solve our problem is applying the idea of independent events. Two (or more) events are said to be *independent* if the probability of their co-occurrence is equal to the product of their individual probabilities.

Once again, this is simple to understand with a nice, easy coin-flipping example. If you flip a coin twice, the probability of getting *heads* twice is equal to the probability of heads multiplied by the probability of heads ($0.5 \times 0.5 = 0.25$), because the outcome of one coin flip has no influence on the outcome of the other (hence, they are *independent* events).

So, when you flip a coin twice, one out of every four times the results will be two heads in a row. If you wanted to know the probability of flipping three heads in a row, the answer is 0.125 ($0.5 \times 0.5 \times 0.5$), which means that three heads in a row happens only one time out of every eight.

In our birthday problem, every time we add another person to the group, we've added another independent event (since one person's birthday doesn't influence anyone else's birthday), and thus we'll be able to figure out the probability of at least two of those people sharing a birthday, regardless of how many people we add; we'll just keep on multiplying probabilities together.

To review, no matter how many people are in our group, only one of two *mutually exclusive* events can occur: at least two people share a birthday or no one shares a birthday. Because of the Law of Total Probability, we know that we can determine the probability of no one sharing a birthday, and one minus that value will be equal to the probability that at least two share a birthday. Lastly, we also know that each person's birthday is *independent* of the other group members. Got all that? Good, let's proceed!

Solving the Birthday Problem

We've already determined that the probability of two people not sharing a birthday in a group of two is equal to 0.997. Let's say we add another person to the group. What is the probability of no one sharing a birthday? There are 363 other birthdays person #3 could have that would result in none of them sharing a birthday. The probability of person #3 not sharing a birthday with the other two is therefore 363/365, or 0.995 (slightly lower).

But remember, we're interested in the probability that *no one* shares a birthday, so we use the rule of independent events and multiply the probability that the first two won't share a birthday by the probability that the third person won't share a birthday with the other two: $0.997 \times 0.995 = 0.992$. So, in a group of three people, the probability that none of them share a birthday is 0.992, which means that the probability that at least two of them share a birthday is 0.008 $(1 - 0.992)$.

This means that only 8 out of every 1,000 randomly selected groups of 3 people will result in at least 2 of them sharing a birthday. This is still a pretty small chance, but note that the probability has more than doubled by moving from two people to three (0.003 compared to 0.008)!

Once we start adding more and more people to our group, the probability of at least two people sharing a birthday starts to increase very quickly. By the time our group of people is up to 10, the probability of at least 2 sharing a birthday is up to 0.117. How do we determine this in general? For every person added to the group, another fraction is multiplied by the previous product. Each additional fraction will have 365 as the denominator, and the numerator will be 365 minus the number of additional people beyond the first.

So, for our previously mentioned group of 10 people, the numerator for the last fraction is 356 $(365 - 9)$, determined like so:

$$\frac{364}{365} \times \frac{363}{365} \times \frac{362}{365} \times \frac{361}{365} \times \frac{360}{365} \times \frac{359}{365} \times \frac{358}{365} \times \frac{357}{365} \times \frac{356}{365} = 0.883$$

This tells us that the probability of no one sharing a birthday in a group of 10 people is equal to 0.883 (much lower than what we saw for 2 or 3 people), so the probability that at least 2 of them will share a birthday is 0.117 $(1 - 0.883)$.

The first fraction is the probability that the second person won't share a birthday with the first person. The second fraction is the probability that the third person won't share a birthday with the first two. The third fraction is the probability that the fourth person won't share a birthday with the first three, and so on. The ninth and final fraction is the probability that the tenth person won't share a birthday with any of the other nine.

In order for no one to share a birthday, every single one of the events in the chain has to co-occur, so we determine the probability of all of them happening in the same group of people by multiplying all of the individual probabilities together. Every time we add another person, we include another fraction into the equation, which makes the final product get smaller and smaller.

Solving for Almost Any Group Size

As the group size increases, it becomes increasingly more likely that at least two people will share a birthday. This makes perfect sense, but what surprises most people is how quickly the probability increases as the group gets bigger. Figure 4-3 illustrates the rate at which the probability goes up when you add more and more people.

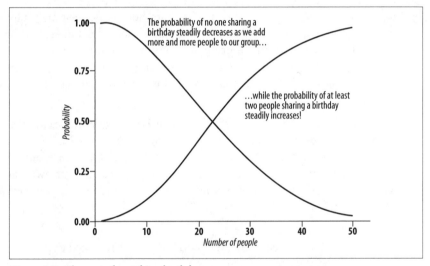

Figure 4-3. Chances of matching birthdays

For 20 people, the probability is 0.411; for 30 people, it's 0.706 (meaning that 7 times out of 10 you will win money on your bet, which are pretty good odds). If you have 23 people in your group, the chances are just slightly better than 50/50 that at least 2 people will share a birthday (the probability is equal to 0.507).

When all is said and done, this is a pretty neat trick that never ceases to surprise people. But remember to make the bar bet only if you have at least 23 people in the room (and you're willing to accept 50/50 odds). It works even better with more people, because your chances of winning go up dramatically every time another person is added. To have a better than 90 percent chance of winning your bet, you'll need 41 people in the room (probability of at least 2 people sharing a birthday = 0.903). With 50 people, there's a 97 percent chance you'll win your money. Once you have 60 people or more, you are practically guaranteed to have at least 2 people in the room who share a birthday and, of course, if you have 366 people present, there is a 100 percent chance of at least 2 people sharing a birthday. Those are great odds if you can get someone to take the bet!

—*William Skorupski*

HACK #46 Design Your Own Bar Bet

With a few calculations, and perhaps some spreadsheet software, you can figure the probabilities associated with all sorts of "spontaneous" friendly wagers.

Several of the statistics hacks elsewhere in this chapter use decks of cards [Hack #42] or dice [Hack #43] as props to demonstrate how some seemingly rare and unusual outcomes are fairly common. As someone who's interested in educating the world on statistical principles, you no doubt will wish to use these teaching examples to impress and instruct others. Hey, if you happen to win a little money along the way, that's just one of the benefits of a teacher's life.

But there's no need to rely on the specific examples provided here, or even to carry cards and dice around (though, knowing you the way I think I do, you might have plenty of other reasons to carry cards and dice around). Here are a couple of basic principles you can use to make up your own bar bet with any known distribution of data, such as the alphabet, numbers from 1 to 100, and so on:

Principle 1

An unlikely event increases in likelihood if there are repeated opportunities for it to occur.

Principle 2

If there are a large number of possible events, the chance of any specific event occurring seems small.

The rest of this hack will show you how to use these principles to your advantage in your own custom-made bar bets.

Principle 1

The probability of any given event occurring is equal to the number of outcomes, which equal the event divided by the number of possible outcomes.

For example, what are the chances that you and I were born in the same month? Pretending for a second that births are distributed equally across all months, the probability is 1/12. There is only one outcome that counts as a match (your birth month), and there are 12 possible outcomes (the 12 months of the year).

What about the probability that any one of *two* people reading this book has the same birth month as me? Intuitively, that should be a bit more likely than 1 out of 12. The formula to figure this out is not quite as simple as one would like, unfortunately. It is not 1/12 times itself, for example. That would produce a smaller probability than we began with (i.e., 1/24). Nor is the formula 1/12 + 1/12. Though 2/12 seems to have promise as the right answer—because it is bigger than 1/12, indicating a greater likelihood than before—these sorts of probabilities are not additive. To prove to yourself that simply adding the two fractions together won't work, imagine that you had 12 people in the problem. The chance of finding a match with my birth month among the 12 is obviously not 12/12, because that would guarantee a match.

The actual formula for computing the chances of an event occurring across multiple opportunities is based on the notion of taking the proportional chance that an event will *not* happen and multiplying that proportion by itself for each additional "roll of the dice." At the conclusion of that process, subtracting the result from 1.0 should give the chance that the event will happen.

This formula has a theoretical appeal because it is logically equivalent to the more intuitive methods (it uses the same information). It is appealing mathematically, too, because the final estimate is bigger than the value associated with a single occurrence, which is what our intuition believes ought to be the case. Think about it this way: how many times will it not happen, and among *those* times, how many times will it not happen on the next occurrence?

Here's the equation to compute the probability that someone among two readers will have the same birth month as I do:

$$1 - \left(\frac{11}{12} \times \frac{11}{12}\right) = 1 - (.917 \times .917) = 1 - .841 = .159$$

Principle 2

To get someone to accept a wager or to amaze an audience with the occurrence of any given outcome, the likelihood must sound small. So, wagers or magic tricks having to do with the 365 days in a year, or the 52 cards in a deck, or all the possible phone numbers in a phone book are more effective and astounding because those numbers seem big in comparison to the number of winning outcomes (e.g., one).

The chance of any unlikely event occurring on any single event is indeed small, so the intuitive belief expressed in this principle is correct. As we have seen, though, the chances of the event occurring increases if you get more than one shot at it, and it can increase rapidly.

Rolling Your Own Bar Bet

Let's walk through the steps that verify my advantage for a couple of wagers I just made up.

Letters of the alphabet. For this wager, I'll pick five letters of the alphabet. I bet that if I choose six people and ask them to randomly pick any single letter, one or more of them will match one of my five letters. Here's how the bet plays out:

Number of possible choices
There are 26 letters in the alphabet.

Probability of a single attempt failing
There are 21 out of 26 possibilities that are not matches: $21/26 = .808$.

Number of attempts
6

Probability of all 6 attempts failing
$.808^6 = .278$

Probability of something other than the previous options occurring
$1 - .278 = .722$

The chance of my winning this bet is 72 percent.

Pick a number, any number. This time, I'll pick 10 numbers from 1 to 100. I bet that if I choose 10 people and ask them to randomly pick any single number from 1 to 100, one or more of them will match one of my ten numbers. Here's how this one works out:

Number of possible choices
There are 100 numbers to choose from.

Probability of a single attempt failing
There are 90 out of 100 possibilities that are not matches: $90/100 = .90$.

Number of attempts
10

Probability of all 10 attempts failing
    $90^{10} = .349$

Probability of something other than the previous options occurring
    $1 - .349 = .651$

The chance of my winning this bet is 65 percent.

On your own. Copy the steps and calculations just shown to develop your own original party tricks. None of these demonstrations require any props, just a willing and honest volunteer.

Notice that the calculations are based on people randomly picking numbers. In reality, of course, people will not pick a letter or number that they have just heard someone else pick. In other words, their choices will not be independent of other choices. If the choices are made based on the knowledge that previous answers are not correct, this helps your odds a little bit. For example, on the 10-out-of-100 numbers wager, if there is no chance that the 10 people will choose a number that has already been chosen, your chances of getting a match go from 65 percent to 67 percent.

Make Sure the Sucker Isn't You!

It is fun to play with others, but you never know when you will get caught in someone else's clever statistics trap. For instance, remember that 1-out-of-12 chance that you have the same birth month as me? I fooled you! I was born in February. There are fewer days in that month than the others, so your chances of being born in that month are actually less than 1 out of 12. There are 28.25 days in February (an occasional February 29 accounts for the .25) and 365.25 days in the year (the occasional Leap Year accounted for again). The chance that you were born in the same month as me is 28.25/365.25, or 7.73 percent, not the 8.33 percent that is 1 out of 12.

So, you are less likely to have the same birth month as me. Come to think of it, the records of my birth, my birth certificate, and so on were lost in a fire many years ago. So, the original data about my birth is now missing.

For all I know, I might not even be born yet!

HACK
#47

Go Crazy with Wild Cards

Wild cards are added to a poker game to ratchet up the fun. Statistically, though, they make things all discombobulated.

Hundreds of years ago, poker players settled on a rank order of hands and decided what would beat what. Pleasantly, for the field of statistics, the order they settled on is a perfect match with the probability that a player will be dealt each hand. Presumably, the developers of poker rules either did the calculations or referenced their own experience as to how frequently they saw each kind of hand in actual play. It is also possible that they took a deck of cards, paper and pencil, and a free afternoon, dealt themselves many thousands of random poker hands, and collected the data. Whatever the method, the rank order of poker hands is a perfect match with the relative scarcity of being dealt those particular combinations of cards.

Rank ordering, though, does not take into account the meaningful distance between one type of hand and the type of hand ranked immediately below it. A straight flush, for example, is 16 times less likely to occur than the hand ranked immediately below it, which is four of a kind, while a flush is only half as likely as a straight, the hand ranked immediately below a flush.

Before we talk about the problem with playing with *wild cards* (cards, often jokers, that can take on any value the holder wishes), let's review the ranking of poker hands. Table 4-17 shows the probability that a given hand will occur in any random five cards, as well as each hand's relative rarity when compared to the hand ranked just below it in the table.

Table 4-17. Poker hands, probabilities, and comparisons

Hand	Probability	Relative rarity
Straight flush	.000015	16 times less likely
Four of a kind	.00024	5.8 times less likely
Full house	.0014	1.4 times less likely
Flush	.0019	2.1 times less likely
Straight	.0039	4.4 times less likely
Three of a kind	.021	2.3 times less likely
Two pair	.048	8.8 times less likely
One pair	.42	1.2 times less likely
Nothing	.50	-----

To gamblers, there are several observations of note from Table 4-17. First, with five cards, half the time players have nothing. Almost half the time, a player has a pair. A player will have something better than a pair only 8 percent of the time.

Second, some hands treated as if they are wildly different in rarity are almost equally likely to occur. Notice that a flush and a full house occur with about the same frequency.

Finally, after three of a kind, the likelihood of a better hand occurring drops quickly. In fact, there are two giant drops in probability: having either nothing or a pair occurs most of the time (92 percent), then two pair or three of a kind occurs another 7 percent of the time, and something better than three of a kind is seen less than 1 percent of the time.

The Problem with Wild Cards

This is all very interesting, but what does it have to do with the use of wild cards? Well, adding wild cards to the deck screws up all of these time-tested probabilities. Assuming that the holder of a wild card wishes to make the best hand possible, and also assuming that one wild card, a joker, has been added to the deck, Table 4-18 shows the new probabilities, compared to the traditional ones.

Table 4-18. Probability of poker hands with one wild card in the deck

Hand	Probability with wild card	Classic probability	Change in probability with wild card
Five of a kind	.0000045	-----	-----
Straight flush	.000064	.000015	+327 percent
Four of a kind	.0011	.00024	+358 percent
Full house	.0023	.0014	+64 percent
Flush	.0027	.0019	+42 percent
Straight	.0072	.0039	+85 percent
Three of a kind	.048	.021	+129 percent
Two pair	.043	.048	−10 percent
One pair	.44	.42	+5 percent
Nothing	.45	.50	−10 percent

The problem with wild cards is apparent as we look at the new probabilities, especially when we look at three of a kind and two pair. Three of a kind is now more common than two pair!

The rank order that traditionally determines which hand beats what is no longer consistent with actual probabilities. Additionally, the chances of getting two pair actually drop when a wild card is added. Other probabilities change, of course, with all the other playable hands becoming more likely. Some super hands, while remaining rare, increase their frequency quite dramatically: hands better than three of a kind are about twice as common as they were before.

Knowing these new probabilities gives smart poker players an edge. In fact, contrary to the stereotype that experienced and professional poker players avoid games with wild cards because they are childish or for amateurs, some informed players seek out these games because they believe they have the advantage over your more naïve types. (You know, those naïve types, like people who don't read Hacks books?)

Why It Works

As you can see in Table 4-18, using wild cards lessens the chance of getting two pair. But why would this be? Surely adding a wild card means that sometimes I can turn a one-pair hand into a two-pair hand. This is true, but why would I? Imagine a player has one pair in her hand, and she gets a wild card as her fifth card. Yes, she *could* match that wild card up with a singleton and call it a pair, declaring a hand with two pairs. On the other hand, it would be smarter for her to match it up with the pair she already has and declare three of a kind. Given the option between two pair and three of a kind, everyone would choose the stronger hand.

The Other Problem with Wild Cards

The existence of wild cards creates a paradox that drives game theorists crazy. The paradox works like this:

1. The ranking of hands and their relative value in a poker game should be based on the frequency of their occurrence. The less frequently occurring hand should be valued more than more commonly occurring hands.

2. In the case of choosing whether to use a wild card to turn a hand into two pair or three of a kind, players will usually choose to create three of a kind. This changes the frequency in practice such that two pair becomes less common than three of a kind.

3. Because rankings should be based on probabilities, the rules of poker should be changed when wild cards are in play to make two pair more valuable than three of a kind.

4. With revised rankings, three of a kind would be worth less than two pair, so now smart players would use their wild card to make two pair instead of three of kind, so two pair would quickly become more common than three of a kind.

5. The ranking rules would then have to be changed again to match the actual frequencies resulting from the previous rule change, and a never-ending cycle would begin.

Table 4-18 avoids this paradox by assuming that players want to make their best hand based on traditional rankings. Clever of me, huh? Want to play cards?

Never Trust an Honest Coin

Of all that is sacred in the often secular world of statistics, no concept has more faith than the honest spin of an honest coin. Fifty percent chance of either heads or tails, right? The troubling answer is, apparently...no!

A basic explanation of chance and how it operates almost always includes a simple example of flipping or spinning a coin. "Heads you win; tails I win" is the customary method for settling a variety of disputes, and the binomial distribution [Hack #66] is usually described and taught as the pattern of random coin outcomes.

But as it turns out, if you spin a coin, especially a brand-new coin, it might land tails up more often than heads up.

Shiny New Pennies

You know the look and feel of a brand-new, mint-condition penny? It's so bright that it looks fake. It's so detailed and sharp around the edges that you have to be careful not to cut yourself.

Well, get yourself one of them bright, sharp little fellas and spin it 100 times or so. Collect data on the heads-or-tails results, and prepare to be amazed because tails is likely to come up more than 50 times. If our understanding of the fairness of coins is correct, a coin should come up tails more than half the time less than half the time. (Say that last sentence out loud and it makes more sense.) Not with the spin of a new penny, though.

New coins, at least new pennies, tend to have a crisp edge that actually is a bit longer or taller on the tail's side (the tail side is imprinted a little deeper into the penny than the head side). Figure 4-4 gives a sense of how this edge looks. If you spin an object shaped like this, there is a tendency for the side with the extra long edge to land face-up.

Imagine spinning the cap from a bottle of beer or soda pop. Not only would it not spin so well, but you also wouldn't be surprised to see it land with the edge side up. A new penny is shaped kind of like a bottle cap, just not quite so asymmetrical. The little extra edge, though, is enough over many spins to give tails the advantage.

Binomial Expectations

The possible existence of a *bottle-cap effect* presents a testable hypothesis:

> The probability of a freshly minted spinning penny landing with tails up is greater than 50 percent.

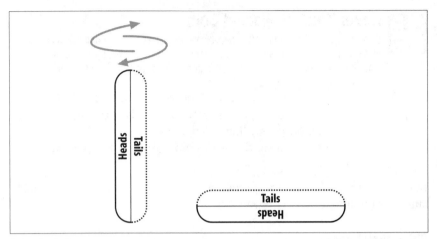

Figure 4-4. Spinning a new penny

Of course, just by chance, over a few flips we might find a coin landing tails up more often than heads, but that wouldn't really prove anything. We know that chance will bring results in small samples that don't represent the nature of the population from which the samples were drawn.

Our sample of coin spins should represent a population of infinite coin spins. If we spin a coin 100 times and find 51 tails, is that acceptable evidence for our hypothesis? Probably not; chance could be the explanation for a proportion other than .50. How about 52 tails? How about 52 percent tails and a million spins?

Statistics comes to the rescue once again and provides a standard by which to judge the outcome of our experiment. We know from the binomial distribution that 100 spins of a theoretically fair coin (one without the unbalanced edge weirdness) will produce 51 or more tails 42 percent of the time. Old-school statistical procedures require that an outcome must have a 5 percent or lower chance of occurring to be treated as *statistically significant*—not likely to have occurred by chance. So, we probably wouldn't accept 51 percent after 100 spins as acceptable support for the hypothesis.

On the other hand, if we spun that hard-working coin 6,774 times and got 51 percent tails, that would happen by chance only 5 percent of the time. Our level of significance for that result is .05. Table 4-19 shows the likelihood of getting a certain proportion of tails just by chance, when the expected outcome is 50 percent tails. Deviations from this expected proportion that are statistically significant can be treated as evidence of support for our hypothesis.

Table 4-19. Coin spins and probability of certain outcomes

Number of spins	Proportion of tails	Probability of the given proportion or higher
100	.51	.42
100	.55	.16
100	.58	.05
500	.51	.33
500	.55	.01
500	.58	.0002
1,000	.51	.26
1,000	.55	.001
1,000	.58	.0000002

Notice that the power of this analysis really increases as the sample size gets big [Hack #8]. You need only a slight fluctuation from the expected to support your hypothesis if you spin that coin 500 or 1,000 times. With 100 spins, you need to see a proportion of tails at or above .58 to believe that there really is an advantage for tails with a newly minted penny.

The distance of the observed proportion from the expected proportion is expressed as a z score [Hack #26]. Here's the equation that produces z scores and generated the data in Table 4-19:

$$z = \frac{\text{Observed Proportion} - \text{Expected Proportion}}{\sqrt{\frac{\text{Expected Proportion}(1 - \text{Expected Proportion})}{\text{Sample Size}}}}$$

The probability assigned is the area under the normal curve, which remains above that z score.

Where It Doesn't Work

Once you prove to yourself that this tail advantage is real, heed this reminder before you go running off to win all sorts of crazy wagers. You must *spin* the coin! Don't *flip* it. Say it with me: *spin, don't flip.*

See Also

- The term *bottle-cap effect* was suggested on an entertaining web page that includes a nice discussion of the extra-tall edge on penny tails. It is maintained by Dr. Gary Ramseyer at *http://www.ilstu.edu/~gcramsey/*.

HACK #49   Know Your Limit

Humans don't always make rational decisions. Even smart gamblers will sometimes refuse a wager when the expected payoff could be huge and the odds are fair. The St. Petersburg Paradox gives an example of a perfectly fair gambling game that perfectly healthy statisticians probably wouldn't play, just because they happen to be human.

The standard decision-making process for statistically savvy gamblers involves figuring the average payoff for a hypothetical wager and the cost to play, and then determining whether they are likely to break even or, better yet, make a boatload of money. Though one could produce dozens of statistical analyses of gambling all about when a person should and shouldn't play, the psychology of the human mind sometimes takes over, and people will refuse to take a wager because it just doesn't feel right.

The Game of St. Petersburg

The game of St. Petersburg is about 300 years old. The parameters of the game were described by Daniel Bernoulli in 1738. Here are the rules:

1. You pay me a fee to play upfront.

2. Flip a coin. If it comes up heads, you win and I'll pay you $2.

3. If it doesn't come up heads, we'll flip again. If heads comes up that time, I'll pay you 2^2 ($4).

4. Supposing heads still hasn't come up, we flip again. Heads on this third flip, and I pay you 2^3 ($8).

So far, it sounds pretty good and more than fair for you. But it gets better. We keep flipping *until* heads comes up. When it eventually arrives, I pay you 2^n, where n is the number of flips it took to get heads.

Great game, at least from your perspective. But here's the killer question: how much would you pay to play?

> The game of St. Petersburg might not really have ever existed as a popular gambling game in the streets of old-time Russia, but it's been used as a hypothetical example of how the mind processes probability when money is involved. It provided many early statisticians an excuse to analyze the way "expected outcomes" works in our heads. The paper was actually published, by the way, by St. Petersburg Academy, thus the name.

Deciding how much you would pay to play is an interesting process. As a smart statistician, you would certainly pay anything less than $2. Even without all the bigger payoff possibilities, betting you will get heads on a coin flip and getting paid more than the cost of playing is clearly a great bet, and you'd go for it in a shot.

You also probably would gladly pay a full $2. You will win the $2 back half the time, and the other half of the time you will get much more than that! This is a game you are guaranteed to win eventually, so it's not a question of winning. When you don't get heads the first time, you have guaranteed yourself at least $4 back, and possibly more—possibly much more.

So, maybe you'd pay $4 to play. Of course, occasionally, your payoff would be really big money—$8, $16, $32, $64...theoretically, the payoff could be close to infinite. But how much would you pay? That's the 64-dollar question.

Statistical Analysis

Some social science researchers suggest that most people would play this game for something around four bucks, maybe a little more. Few would pay much more. What about statistically, though? What is the most you *should* pay?

Well, this is where I consider turning in my Stats Fan Club membership card, because I am afraid to tell you the correct answer. The rules of probability as they relate to gambling suggest that people should play this game at any cost. Yes, a statistician would tell you to play this game for *any price*! As long as the cost is something short of infinity, this is, theoretically, a good wager.

Let's figure this out. Here's the payoff for the first six coin flips:

Flips	Likelihood	Proportion of games	Winnings	Expected payoff
1	1 out of 2	.50	$2	$1
2	1 out of 4	.25	$4	$1
3	1 out of 8	.125	$8	$1
4	1 out of 16	.0625	$16	$1
5	1 out of 32	.03125	$32	$1
6	1 out of 64	.015625	$64	$1

Expected payoff is the amount of money you would win on average across all possible outcomes. For a single flip, there are two outcomes: for heads, you win $2; for the other possibility, tails, you get $0. The average payout is $1, the expected payoff for one coin flip (and, it turns out, for any number of coin flips).

If you play this game 64 times, you will get to the sixth coin flip just once, but you will win $64. 32 of those 64 times you will win just $2. The average payoff sounds low—just a buck. Occasionally, though, heads won't come up for a very long time, and when it finally does, you have won yourself a lot of money. When you start the game, you have no idea how long it will go and it could be very long indeed (a lot like a Peter Jackson film).

Notice a few things about this series of flips and how the chances drop at the same rate as the winnings go up:

- Only six coin flips are shown. Theoretically, the flipping could go on forever, though, and no head might ever come up.

- With each coin flip, the winnings amount continues to double and the proportion of games where that number of flips would be reached continues to be cut in half.

- The "Proportion of games" column never adds to 1.0 or 100 percent, because there is always some chance, no matter how very small, that one more flip will be needed.

The decision rule among us Stats Fan Club members for whether to play a gambling game is whether the *expected value* of the game is more than the cost of playing. Expected value is calculated by adding up the expected payoff for all possible outcomes.

You'll recall that the expected payoff for each possible trial is $1. There are an infinite number of possible outcomes, because that coin could just keep flipping forever. To get the expected value, we sum this infinite series of $1 and get a huge total. The *expected value* for this game is infinite dollars. Since you should play any game where the cost of playing is less than the expected value, you should play this game for any amount of money less than infinity.

Why It Doesn't Work

Of course, in real life, people won't pay much more than $2 for such a game, even if they knew all the statistics. No one really knows for sure why smart people turn their noses up at paying very much money for such a prospect, but here are some theories.

Infinite is a lot. Even if you accept in spirit that the game is fair over the long run and would occasionally pay off really big if you played it many, many times, that "long run" is infinitely long, which is an awfully long time. Few people have the patience or deep enough pockets to play a game that relies on so much patience and demands such a large fee.

Decreasing marginal utility. The originator of the problem, Bernoulli, believed that people perceive money as valuable, but the perception is not proportional to the amount of money. In other words, while having $16 is better than having $8, the relative value of one to the other is different than the relative value of having $128 compared to $64.

So, at some point, the infinite doubling of money stops being equally meaningful as a prize. Bernoulli also believed that if you have a lot of money, a small wager is less meaningful than if you have very little money. (Kind of like those wealthy cartoon characters who light their cigars with hundred dollar bills.)

Risk versus reward. Humans tend to be risk averse. That is, they will occasionally risk something in exchange for a reward, but they want that risk to be fairly close to the chances of success. It is true that the game of St. Petersburg has a chance for a massive reward, but the chance might be seen as too little compared to a risk of even $4.

Infinity doesn't exist. Some philosophers would argue that people do not accept the concept of infinity as a concrete reality. Any sales pitch to encourage people to play this game by promoting the infinity aspects would be less than compelling.

This might be why I don't buy lottery tickets. I don't play the lottery because my odds of winning are increased only slightly by actually playing. In my mind, the odds of me winning are infinitely small, or close enough to it that I don't treat the possibility of winning as real.

See Also

- "Gamble Smart" [Hack #35]
- A very interesting and thoughtful discussion of the St. Petersburg Paradox is in the *Stanford Encyclopedia of Philosophy*. The online entry can be found at *http://plato.stanford.edu/entries/paradox-stpetersburg*.

Playing Games
Hacks 50–60

Games don't have to involve gambling to involve statistics. You can use knowledge of game-specific probabilities to win on TV game shows [Hack #50], at Monopoly [Hack #51], or when coaching a football team [Hack #58].

The most common place you see statistics in your everyday life is probably in the world of sports, though the word "statistics" isn't really used the same way a stat-hacker uses it. Sports fans tend to think of the *data* as the *statistic*. Regardless, there are plenty of hacks that can help you predict the outcome of a game before it is over [Hack #56] or even begun [Hack #55].

Since history is always our best guide to the future, your best predictions will require various ways to track, visualize [Hack #57], and rank [Hack #59] the performance of teams and players.

Of course, if you have the heart of a true statistics hacker, then you think that some statistical games—such as building a learning computer out of coconuts [Hack #52], doing card tricks through the mail [Hack #53], keeping your iPod honest [Hack #54], or estimating the value of pi purely by chance [Hack #60]—are fun all by themselves.

HACK #50 Avoid the Zonk

On the TV game show Let's Make a Deal, contestants often had to choose between three curtains. For these sorts of situations, there is a statistical strategy that will help you to win the Buick instead of the lifetime supply of Rice-A-Roni.

Imagine, if you will, that you are traveling with your Uncle Frank through an uncharted region of Tonganoxie, Kansas. You come to a fork in the road that branches out into three possible paths: A, B, and C. You don't know which will lead you to your destination, the fabled world's largest ball of twine (in Cawker City, Kansas). An old prospector is resting with his burro at the crossroads.

"Say, old timer," you say, "which road leads to the world's largest ball of twine?"

"Well," says he, "I know, but I won't tell you. What I will do, though, is tell you that one road is the correct road. Two are wrong and lead to certain disaster (or at least poorly maintained restrooms). Go ahead and take your pick, city slicker. As you drive off, look back at me. I *won't* signal whether you are right or wrong, but I will point at *one* of the other two roads. The one I point at will be a wrong road. You still won't know for sure whether you guessed right or not, of course, but I guarantee that I'll point at one of the two roads you are not on and it will be a wrong road."

You accept the strange man's offer (what choice do you really have?) and you ask Uncle Frank, the experienced gambler among you, to pick a road. He does so randomly and you head off optimistically down one of the three paths—let's say A. As you look back, the kindly prospector points to one of the other roads—let's say B. Immediately, you slam on the brakes and back the car up. Over the objections of Uncle Frank, you head down the remaining road, C, with the peddle to the metal, fairly confident that you are now on the right path.

Crazy, are you? Suffering from white-line fever? No, you've just applied the statistical solution to what is known as *the Monty Hall problem* and chosen the road among the three that has the greatest chance of being correct. Hard to believe? Read on, my friend, and prepare to win riches beyond your wildest dreams.

The best strategy in this case is so counterintuitive and downright weird that the world's smartest people have disagreed aggressively about whether it even really is the best strategy. But believe me—it is.

The Monty Hall Problem and Game Show Strategy

In our example with the three roads and the prospector, there is, in fact, a two-thirds (about 67 percent) chance that C is the correct road. To apply this odd strategy to a more realistic situation, think of contestants on game shows or gamblers in any game in which prizes are hidden in boxes or behind doors. As typically discussed among game show theorists and cranky statisticians, the problem is presented as a fairly common actual situation on the game show *Let's Make a Deal* (which had its heyday in the 1960s and 1970s), but it is a situation still seen today in TV game shows. The host of *Let's Make a Deal* was Monty Hall, so the problem carries his name.

As a game show scenario, the problem goes like this. Monty presents to you three curtains. He knows what is behind each curtain. He explains that behind one of the curtains is a brand-new car. The other two curtains hide worthless prizes, what Monty used to call *zonks*. (Zonks were often something like a donkey or a giant rocking chair, something that wouldn't be of any real use.) He lets you pick a curtain, and you will win whatever is behind it. Let's say you pick curtain A. He then opens one of the unchosen curtains—B, for example—to show you that it has a zonk behind it. He then offers to let you trade your original choice for the remaining curtain, C. Should you switch?

As with the three roads problem, the answer is yes, you should switch. The answer just never seems right the first time one hears it. But, if you want to increase your odds of winning the car, you should now switch.

Why You Should Always Switch

Think of the probability of you guessing the correct curtain. Let's assume that it is a random guess—none of this "I notice that one curtain moved, so I figured there was a donkey behind it" stuff.

Three curtains, with only one curtain being a winner, means there is a 1 out of 3 chance that you will guess right and win the car. That's about 33 percent. On that first guess, with no additional information, you are likely to be wrong; in fact, you have a 2 out of 3 chance of being wrong. In other words, there is about a 67 percent chance that the car is somewhere behind the two curtains you did not pick.

Once you know that one of those other two curtains does not have the car, that doesn't change the original probability that the car is 67 percent likely to be somewhere behind those two unselected curtains. Remember, Monty will always have a wrong curtain he can open, no matter which one you choose. The 67 percent chance that the car is behind B or C remains true, even after B is revealed to not be hiding the car. The 67 percent likelihood now transfers to curtain C. That's why you should always switch to the other curtain.

 If you were given the option of swapping your pick of one curtain for *both* the other two curtains, you'd switch in a second wouldn't you? That's essentially what is offered in the Monty Hall problem.

Some figures might be necessary to persuade your inner skeptic. Look at Table 5-1, which shows the probability breakdown for the three options at the start of the game. You have a one-third chance of guessing the winning curtain and a two-thirds chance of picking a nonwinning curtain.

Table 5-1. Probability of car's location at start of game

Curtain A	Curtain B	Curtain C
33.33 percent	33.33 percent	33.33 percent

Table 5-2 shows the same probabilities grouped in a different way, but it hasn't changed any of the parameters of the problem.

Table 5-2. Restated probability of car's location at start of game

Curtain A	Curtain B or Curtain C
33.33 percent	66.66 percent

Table 5-3 shows the probabilities after Monty reveals one of the nonchosen curtains (Curtain B) to be a nonwinner. The 67 percent likelihood now transfers to curtain C.

Table 5-3. Probability of car's location after curtain B is opened

Curtain A	Curtain B	Curtain C
33.33 percent	0.00 percent	66.66 percent

In any situation like this, you should switch. You might be wrong, of course, but you have a better shot of winning that car or whatever other prize you are playing for if you accept any offers to switch. This is always the best strategy, if a few criteria are met:

- The host knows what is behind each curtain.
- The host reveals one of the unchosen curtains and the prize is not behind it.
- Your original choice was random.

Don't be too concerned if the correctness of this solution isn't immediately apparent. Really smart people often first view the new odds as being 50/50 between the two unopened curtains and, therefore, it doesn't matter if you switch. The key to remember, though, is that your original chance of picking the correct door, 33.3 percent, cannot change no matter what happens after you make your choice. Even experts sometimes disagree about the best way to view this question. Even people as wise as the old prospector you met out in Tonganoxie that started our discussion don't always know the right answer to the Monty Hall problem. How do you think he won that burro?

The Controversy

The Monty Hall problem and the general game show strategy that resulted was first introduced to the masses in 1991 by Marilyn Vos Savant, a columnist for *Parade Magazine*. Because she is known for being a "high IQ genius," Vos Savant answered questions from readers, sometimes of a brain teaser nature. Someone sent in the problem as I've described it, and she published the answer I have given here.

Apparently, she received many letters, some angry, from statisticians, philosophers, and such claiming that she got it wrong. In scholarly journals, there were even published debates about whether her answer was correct. My read of the debate is that it turned out that most of the arguments centered on a key ingredient of the question: Monty knows what is behind each door, so when he opens that first curtain, he knows it will be a zonk. Otherwise, the reveal does not count as new information and the answer Vos Savant gave does become debatable. Most of the critics of her answer missed that part of the original published question.

HACK #51 Pass Go, Collect $200, Win the Game

Monopoly is a game of chance (and Chance cards). As such, the best strategies for winning capitalize on probability.

Winning the popular Parker Brothers board game *Monopoly* requires negotiating skill, clever money management, and insightful investment planning. It also requires a little bit of luck.

As two six-sided dice (and a randomly shuffled pile of cards) are the primary determinants for deciding what square you land on, luck pays more than just a small role in the outcome. Competitive statisticians such as you and me (or, at least, me) are drawn to any game in which probability plays a key part because, by applying a few probability basics, we should win more often than your average, run-of-the-mill railroad baron.

Monopoly Statistical Basics

Let's start by examining the simple effects of rolling two dice. Figure 5-1 shows the most common squares landed on in the first couple of turns for everyone.

Imagine the start of the game, when everybody is on Go. With two six-sided dice, there is a 44.5 percent chance that a 6, 7, or 8 will be rolled, with 7 as the most likely outcome (16.7 percent). For your first two dice rolls, then,

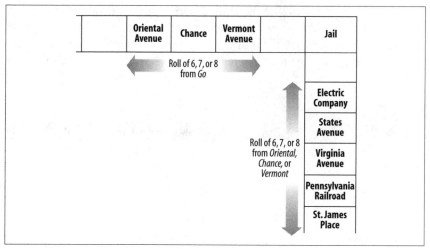

Figure 5-1. Likely opening rolls

some squares are more likely to be hit (e.g., the light blues and Virginia Avenue) and some less likely (Baltic Avenue or Income Tax). Based on opening dice rolls alone, not all squares are equally likely to be landed on.

> Poor Mediterranean Avenue cannot even be landed on when starting at Go, because a dice roll of 1 is not possible with two dice. Have you ever noticed that it is almost always one of the last properties still available for purchase?

The Go square is a good starting point to begin calculating the various likelihoods for landing. Not only does everyone start there at the beginning, but there is also a Chance card that sends players there. On the other hand, if a player hits the "Go to Jail" space, she goes directly to jail, bypassing Go. So, the probability for landing on Go is affected by not just the possible permutations of dice rolls, but also the various Chance cards, which send players various places, and the rules of the game itself, which include squares that make things happen, going to jail situations, and getting out of jail situations.

Key Properties

I've been using Go as an example square, but, of course, Go isn't even a square we can purchase. What we really want to know is what properties to buy or trade for and where to build first. We want high traffic areas; the secret to real estate success is "location, location, location" (and, apparently, for some reason I've never understood, a nice wooden deck).

Table 5-4 shows the top 20 most landed-upon squares, taking all rules into account. The table also shows the chance that a player will come to rest on any one of those squares. Keep in mind that an "average" square has a 2.5 percent chance of being your final resting place (40 squares divided by 100 is 2.5).

Table 5-4. Best real estate in all of Atlantic City

Square	Rank	Chance of ending your turn on it
Jail	1	11.60 percent
Illinois Avenue	2	2.99 percent
Go	3	2.91 percent
B & O Railroad	4	2.89 percent
Free Parking	5	2.83 percent
Tennessee Avenue	6	2.82 percent
New York Avenue	7	2.81 percent
Reading Railroad	8	2.80 percent
St. James Place	9	2.68 percent
Water Works	10	2.65 percent
Pennsylvania Avenue	11	2.64 percent
Kentucky Avenue	12	2.61 percent
Electric Company	13	2.61 percent
Indiana Avenue	14	2.56 percent
St. Charles Place	15	2.56 percent
Atlantic Avenue	16	2.54 percent
Pacific Avenue	17	2.52 percent
Ventnor Avenue	18	2.52 percent
Boardwalk	19	2.48 percent
North Carolina Avenue	20	2.47 percent

Table 5-4 is derived from information provided by Truman Collins on his web site at *http://www.tkcs-collins.com/truman/monopoly/monopoly.shtml*. Clever Mr. Collins developed both probability trees and a computer simulation to verify these values, and offers them for two situations: when players wish to remain in jail as long as possible (to earn rent and not have to pay rent) and when they wish to get out of jail as quickly as possible (to buy still available properties). I reported the values that apply to the former strategy.

You can draw some important tactical conclusions from this data:

Capitalize on the jailbirds

A remarkable 12 percent of the time, your opponent will begin a turn on the Jail square. Clearly, owning and developing the land that recently released parolees are most likely to land upon is a wise goal. This amounts to the orange properties (St. James and his brothers) and, to a lesser extent, the reds (e.g., Illinois Avenue) and the purples (St. Charles and friends).

Own the oranges

All three orange properties are in the top 10. About 1 out of every 12 rolls will result in a hit on Tennessee or New York Avenue or St. James Place. Getting the monopoly with these properties and developing quickly would seem to be the strategy that a pure statistician would choose.

Avoid the far side

Properties on the far side of the board—the greens, Boardwalk, and Park Place—are less likely to be landed upon, even deep into the game. Only Boardwalk and Pacific Avenue rank high, and Boardwalk is there, no doubt, because there is a Chance card that sends players there. These properties are also the most expensive to develop, so including these monopolies prominently in one's game plan is a bit risky.

Importance of the Monopoly Prison System

Without a statistical analysis, it might not be so clear the crucial role that the Jail and "Go to Jail" squares play in the overall true value of real estate. One wishes it was for sale. Players will start or end their turn on the Jail square more often than they will land on any monopoly on the board. A constant stream of released prisoners flood across one side of the board, increasing the opportunity to collect rents on properties all the way up to Illinois.

Jail can also provide a welcome respite from having to travel the streets paying rent to other players, though early in the game, Jail can prevent you from buying up your dream properties. A final observation on the importance of Jail: there is only one square that you can never end your turn on. Can you name it? *Go to Jail.*

See Also

- Bill Butler runs another web site that presents the probabilities associated with Monopoly at *http://www.durangobill.com/Monopoly.html.* Among other things, the site hosts a discussion of the incredible calculation difficulties involved when one wishes to include every real-life detail of Monopoly play, such as keeping track of whether a particular Chance or Community Chest card has been drawn already.

- The basic formula for calculating the probability of landing on a square (with cool London, England, street names in the example) is presented at *http://hometown.aol.co.uk/monopolycheat/prob/method.html*.

Use Random Selection as Artificial Intelligence

#52   Statisticians have been able to build intelligent, learning computers long before the advent of the microprocessor. You can use coconut shells and the laws of probability to build a machine that will learn to never lose at Tic-Tac-Toe.

A common joke about the 1960s TV show *Gilligan's Island* is that the Professor was always building computers or washing machines or rocket ships out of coconuts and vines. I don't know about washing machines and rocket ships—those do sound pretty far-fetched—but the castaways certainly could have built computers out of coconuts. You can, too. If you are ever stranded on a desert island and want a companion, build one.

You won't need a volleyball like Tom Hanks's buddy in *Castaway*, and it won't have much personality, but your computer will be able to play games with you, and it will even learn and get smarter. The driving forces behind the learning algorithm are chance and the power of random selection.

Trial-and-Error Learning

According to behavioral psychologists, all animals (including humans, otters, and single-celled creatures) learn essentially the same way. Experience presents situations in which choices lead to outcomes. As the animal receives feedback about the outcome, it adapts. If the outcome was positive, the creature is more likely to make the same choice in the future. If the outcome was negative, the creature is less likely to make that choice again.

Notice that there is no guarantee that a "good" behavior is always repeated or that a bad behavior becomes extinct; it is only a matter of probability. The right decision is more likely to be made and the wrong decision is less likely to be made. To make a machine that mimics the way that animals learn, we must build on this probability angle.

Game playing reflects much of the trial-and-error learning process because outcomes are easily interpreted as positive (a win) or negative (a loss). In games, the feedback is often immediate, and studies show that the closeness in time between the choice and the feedback is a key factor in whether learning has occurred. And learning, remember, is defined here as an increase in the likelihood of correct choices or a decrease in the likelihood of incorrect choices.

Building a Tic-Tac-Toe Machine

Stuck on this island with no friends, you might want to fight boredom by playing games with a smart opponent. Here are instructions for building a contraption that does not use any electricity or silicon, but will play a game and provide decent competition.

This machine learns: the more times you play against it, the better it will be. The game this machine plays is Tic-Tac-Toe, but theoretically, you could build a device for any two-person strategy game using the same principles. Tic-Tac-Toe is simple enough that it demonstrates well the methods of design, construction, and operation.

If the Professor on *Gilligan's Island* ever did build a computer out of coconuts, he was likely influenced by the pioneering work of biologist Donald Michie and his matchboxes. Michie published an article in the very first issue of the *Computer Journal* in 1963, a few years before Gilligan and his pals were stranded on their island. Michie describes how he designed and actually built a nonelectric computer with the following complete list of parts:

287 matchboxes
> Each matchbox has a little drawer that can be opened. Michie labeled each matchbox with one of 287 different possible Tic-Tac-Toe configurations throughout a game. There are actually many more possible positions, but because the standard Tic-Tac-Toe layout of three rows and three columns is symmetrical, four different unique positions can be summarized with just one position. At any point in the game, the current layout of the "board" directs the human operator to the corresponding matchbox.

A large supply of beads of nine different colors
> The nine colors represent each of the nine different spaces on the Tic-Tac-Toe board. Each matchbox begins with an equal supply of beads for each of the possible next moves. Only beads representing legal moves are put in each box. Different positions and matchboxes, of course, correspond to only a small set of legal next moves, so each box has a slightly different mixture of beads.

The Professor would have used coconut shells instead of matchboxes and sand pebbles or seeds (or perhaps Mr. Howell's coin collection, which he never goes anywhere without) instead of beads. Gather these supplies from your tropical surroundings, organize the pebble-filled coconuts in an efficient grouping, and you have your desert island game-playing computer. Yes, you'll need to find 287 coconuts, but do you have anything better to do?

Operating the Computer

To play a game of Tic-Tac-Toe against your pebble-powered PC, follow these instructions:

1. The computer goes first. Find the coconut that is labeled with the current position. (For the first move, it is a blank layout.) Close your eyes and randomly draw out a pebble.

2. Mark an X on your board (drawn in the sand, I'm assuming) in the space indicated by the color of the pebble. Set the pebble aside in a safe place.

3. Make your move, marking an O in your chosen space.

4. There is a new position on the board now. Go to the corresponding coconut and randomly draw out a pebble from it. Return to step 2.

5. Repeat steps 2 through 4 until there is a winner or a draw.

What happens next is the most important part because it results in the computer learning to play better. Behavioral psychologists call this final stage *reinforcement*.

If the computer loses, "punish" it by taking the pebbles that you drew randomly from the coconuts and throwing them into the ocean.

If the machine wins or draws the game, return the pebbles to the coconuts from which they came and "reward" it by adding an additional pebble of the same color.

Why It Works

The process of rewarding or punishing the computer essentially duplicates the process by which animals learn. Positive results lead to an increase in the likelihood of the rewarded behavior, while negative results lead to a decrease in the likelihood of the punished behavior. By adding or removing pebbles, you are literally increasing or decreasing the true probability of the machine making certain moves in the game.

Consider this stage of a game, where the computer, playing X, must make its move:

X	O	X
	O	

You probably recognize that the best move—really, the only move to consider—is for the computer to block your impending win by putting its X in the bottom center space. The computer, though, recognizes several possibilities. It considers any legal move. Two moves that it would consider (which means, literally, that it would allow to be drawn randomly out of the coconut shell) are the best move and a bad move:

X	0	X
	0	
	X	

X	0	X
	0	
X		

When the computer first starts playing the game, both these moves (or behaviors) are equally likely. Other moves are also possible in this situation, and they are also equally likely. The move on the left probably won't result in a loss, at least not immediately, so as pebbles representing that move are added to the coconut, the relative probability of that move increases compared to other moves. The move on the right probably ends in a loss (except against Gilligan, maybe), so the chance of that move being selected next time mathematically decreases, as there are fewer pebbles of that color to be randomly selected.

The probability of any given move being selected can be represented by this simple expression:

$$\frac{\text{number of pebbles representing the move}}{\substack{\text{total number of pebbles in the coconut} \\ \text{corresponding to current board layout}}}$$

The machine begins with an equal number of pebbles or, in other words, an equal likelihood of any of a variety of moves being chosen. Of course, some moves look foolish to our experienced game-playing eye and would never be made in a real game except by the most naive of players. The point that behavioral psychologists argue, though, is that all creatures are novices until they have built up a large pool of experiences that have shaped the basic probabilities that they will engage in a behavior.

Hacking the Hack

There are several ways to modify your machine to make it smarter. For example, you can choose to reward moves that lead to wins more than moves that lead to ties. This should produce a good player more quickly. Michie suggested three beads for a win and one bead for a tie.

If you want to simulate the way animal learning occurs, you can adjust the system so that moves near the end of the game are more crucial than those made at the beginning. This is meant to mirror the observation that reinforcement that comes closest in time to when the behavior occurs is most effective. In the case of Tic-Tac-Toe, mistakes that lead to immediate losses should be dealt with and punished more effectively. By having fewer total beads in use for moves late in the game, the learning will occur more quickly.

An obvious upgrade is to make your computer smarter by not even allowing bad moves. Don't even place pebbles representing moves that will result in immediate defeat into your containers. This will solve the problem of your computer's initial low intelligence, but it doesn't really reflect the way animals learn. So, while this might make for a stronger competitor, the Professor would be disappointed in your lack of scientific rigor.

HACK #53 Do Card Tricks Through the Mail

A shuffled deck of cards is meant to be random. Scientific analyses show that it actually isn't random, and you can capitalize on known probabilities of card distributions to perform an amazing card trick for people you have never met.

Imagine you receive a thick, mysterious envelope in the mail. Rather than having it disposed of by the nearest domestic security officers, you open it and find an ordinary deck of cards and the following set of instructions:

1. Cut the deck.
2. Shuffle the cards once, using a *riffle* shuffle (defined later in this hack).
3. Cut the deck again.
4. Shuffle the cards one more time using a riffle shuffle.
5. Cut the deck again.
6. Remove the top card of the deck, write it down, and place it anywhere in the deck.
7. Cut the deck again.
8. Shuffle again.
9. Cut one more time.
10. Mail this deck back to the enclosed address (a post office box in Tonganoxie, Kansas, or some other place with a name that conjures up wonder and whimsy).

You follow all these instructions (while wearing protective rubber gloves) and return the deck. About a week later, a smaller envelope arrives. In it is your chosen card! (There also might be a request for $300 and an offer to predict your future, but you just throw the offer away.)

Amazing, yes? Impossible, you say? Thanks to the known likely distribution of shuffled cards, it is more than possible, and even a budding statistician like you can do it. No enrollment in Hogwarts necessary.

How It Works

Quite a bit is known, mathematically, about the effects of various types of shuffles on a deck of cards. Though a thorough shuffle (such as a *dovetail* or *riffle* shuffle, which interlaces two halves of the deck) is meant to really scramble up a deck from whatever order the cards were in to some new order that's quite different from the original, parts of the original sequence of cards remain even after several cuts and shuffles.

Statisticians have analyzed these patterns and published them in scientific journals. The work is similar to that which resulted in the groundbreaking suggestion that one should shuffle a deck of cards exactly seven times to attain the best mix before dealing the next round of hands for poker, spades, or bridge.

Picture a deck of cards in some order. After one shuffle, if the shuffle is perfect, the original order would still be visible within the now supposedly mixed distribution of cards. In fact, there would be two original sequences now overlapping each other, and by taking the alternate cards, you could reconstruct the original overall order.

Table 5-5 shows a deck of cards before and after a perfect shuffle. Just 12 are shown for efficiency's sake, but these principles all apply to a full 52-card deck.

Table 5-5. Effect of perfect shuffling on card distribution

Before shuffle	After perfect riffle shuffle
1. Ace of Clubs	1. Ace of Clubs
2. Two of Clubs	7. Seven of Clubs
3. Three of Clubs	2. Two of Clubs
4. Four of Clubs	8. Eight of Clubs
5. Five of Clubs	3. Three of Clubs
6. Six of Clubs	9. Nine of Clubs
7. Seven of Clubs	4. Four of Clubs
8. Eight of Clubs	10. Ten of Clubs
9. Nine of Clubs	5. Five of Clubs
10. Ten of Clubs	11. Jack of Clubs
11. Jack of Clubs	6. Six of Clubs
12. Queen of Clubs	12. Queen of Clubs

If you knew the starting order of these 12 cards, you could pick it out fairly easily by just looking at every other card in the new grouping. These subpatterns are characterized as *rising sequences*: the cards rise in value as you move along the sequence. If cards begin in one long rising sequence (or a group of four, because there are four suits), riffle shuffles will maintain these rising sequences; they will just be interwoven together. These groupings of rising sequences will remain, even after many shuffles.

If at any time during the shuffling and cutting of the deck, a card is taken from the deck and purposefully placed anywhere else in the deck, it will appear "out of place" compared to the overall pattern of rising sequences. This, of course, is exactly what the card trick's instructions demand, and it explains how your mysterious magician (or you when you assume that role) could spot what card has been moved.

For the sequence shown in Table 5-5, imagine that the Ace of Clubs (#1 in the original sequence) was removed from the top of the deck and placed randomly somewhere in the middle of the cards. Let's say it ends up, between the 4 and 10 of Clubs (between #4 and #10 in the new distribution). It would now be permanently out of sequence, and it is unlikely that anymore shuffling would move it back to where it belongs.

Cutting the cards between shuffles does nothing to affect the overall sequence, if we think of a deck of cards as an endless loop. Nonstandard shuffles, such as cutting the deck into three equal piles and changing the order of those piles before shuffling, will break down the sequence, however, and the magic trick instructions must clearly state that cards should be cut once into two piles.

Of course, realistic analytic work about what happens to playing cards in the hands of real-life people must take into account that people are human and make human errors. As the philosophers say, "To shuffle badly is human." Some cards that should have been separated by exactly one card in a perfect riffle shuffle might, unpredictably, be separated by two cards or might be adjacent to each other and not separated at all. Table 5-6 shows one possible outcome of a more human, less perfect, shuffle.

Table 5-6. Possible effect of sloppy shuffling on card distribution

Before shuffle	After realistically human riffle shuffle
1. Ace of Clubs	1. Ace of Clubs
2. Two of Clubs	7. Seven of Clubs
3. Three of Clubs	8. Eight of Clubs
4. Four of Clubs	2. Two of Clubs
5. Five of Clubs	3. Three of Clubs
6. Six of Clubs	9. Nine of Clubs

Table 5-6. Possible effect of sloppy shuffling on card distribution (continued)

Before shuffle	After realistically human riffle shuffle
7. Seven of Clubs	10. Ten of Clubs
8. Eight of Clubs	5. Five of Clubs
9. Nine of Clubs	4. Four of Clubs
10. Ten of Clubs	11. Jack of Clubs
11. Jack of Clubs	6. Six of Clubs
12. Queen of Clubs	12. Queen of Clubs

This randomness in how a person will actually shuffle the cards creates both a dilemma and an opportunity. The dilemma is that correctly identifying which card is out of sequence is now not certain, because the sequences cannot be perfectly reconstructed and the magician must rely a bit on probabilities, which adds some risk to the trick.

The opportunity comes when the subject of the trick realizes that you could not possibly count on the execution of perfect shuffles. When you identify the chosen card anyway, in the midst of this random uncertainty, the bewilderment will be even greater.

Probability of Success

Because the exact nature of the scrambling of the deck cannot be known, the magician can identify a card as out of sequence only because the shuffles were less than perfect. Also, the trick is much more likely to be successful (only one card is out of sequence) if the instructions do not allow anymore cutting or shuffling after the card is taken from the top of the deck and placed in the middle.

Statisticians from Columbia and Harvard University, Dave Bayer and Persi Diaconis, have conducted a mathematical exploration of the possible outcomes of a deck of cards shuffled and mixed in the ways described for this magic trick. (Presumably, the faculty at these institutions has a lot of free time on its hands?) They developed a mathematical formula for identifying the one card out of place and ran a million computer simulations to test the accuracy of guesses by their cyber-sorcerer as to the chosen card. Their analysis assumed perfect dovetail shuffles. They found that with only a couple of shuffles, the trick works pretty well, but the odds of success decrease quickly as more shuffles are allowed.

Table 5-7 shows the probability of success for a 52-card deck shuffled different numbers of times. It also shows the chances that the correct card would be chosen if more than one guess were allowed.

Table 5-7. Chance of pulling off the seemingly impossible

Number of guesses	Two shuffles	Three shuffles	Four shuffles	Five shuffles	Six shuffles
1	99.7 percent	83.9 percent	28.8 percent	8.8 percent	4.2 percent
2	100 percent	94.3 percent	47.1 percent	16.8 percent	8.3 percent
3	100 percent	96.5 percent	59.0 percent	23.8 percent	12.3 percent

Of course, the odds go down slightly when one takes into account random errors in real-world shuffling, but the relative success rate would still be as Table 5-7 indicates. If you perform the trick as described—one guess, after three shuffles—the guess should be correct around 80 percent of the time (lowering the 83.9 percent estimate somewhat arbitrarily to take into account bad shufflers).

To play it safe, you might do the trick with at least three people. Then, assuming 80 percent likelihood for each person, the chances that you will amaze at least one of those three people increases to 98.4 percent, which is almost a certainty. If you are wrong on all three, just never speak or write to those people again, close your post office box, and concentrate on more important things in life. After all, with hard work, you might get into Columbia or Harvard someday and do really important things.

See Also

- The Bayer and Diaconis study appeared in 1992 in *The Annals of Applied Probability, 2,* 2, 294–313. In that article, they cite two magicians as early developers of card tricks based on the *rising sequences* principle (see the following two bullets).
- Williams, C.O. (1912). "A card reading." *The Magician Monthly, 8,* 67.
- Jordan, C.T. (1916). "Long distance mind reading." *The Sphinx, 15,* 57. This is the presentation on which the effect described in this hack is based.

Check Your iPod's Honesty

HACK #54   Find out how random your iPod's "random" shuffle really is.

Personalized song ratings in Apple's iTunes, the software that allows you to play songs on your iPod, lets you quickly find your favorites and helps the Party Shuffle feature play more of what you like most. The algorithm iTunes uses to pick what comes next in the playlist is meant to select randomly from your favorites. Is it really random, though?

After hearing one artist played over and over during a shuffled play of your entire music library in iTunes, you might think your player has a preference of its own. Apple, though, claims the iTunes's shuffle algorithm is completely random. The shuffle algorithm chooses songs *without replacement*. In other words, much like going through a shuffled deck of cards, you will hear each song only once until you have heard them all (or until you have stopped the player or selected a different playlist).

iTunes's Party Shuffle is a different matter. Its algorithm selects songs *with replacement*, meaning the entire library is reshuffled after each song is played (like reshuffling a deck of cards after every time a card is drawn). The "Play higher rated songs more often" option does exactly what it says, but how much preference is given to higher rated songs?

This hack originally appeared as an article on the OmniNerd web site at *http://www.omninerd.com/*.

Assessing iTunes's Selection Procedures

I wanted to test two different song selection options: *Party Shuffle* and "Play higher rated songs more often." I created a short playlist of six songs: one from each different star rating and a song left unrated. The songs were from the same genre and artist and were each changed to be only one second in duration.

I conducted my tests on iTunes 5. iTunes 6 has added a *Smart Shuffle* feature, which may decrease the chances of hearing songs from the same artist or album consecutively, but I haven't tested it yet.

After resetting the play count to zero, I hit Play and left my desk for the weekend. I ran the same songs twice: once selecting *random* (Party Shuffle) and once selecting both *random* and the "Play higher rated songs more often" option. Table 5-8 shows the play counts, as of Monday morning.

Table 5-8. Song selection distribution

	Random selection		Based on rating	
Song rating	Times played	Percentage of total	Times played	Percentage of total
None	9,105	16.70 percent	2,052	3.9 percent
1	9,055	16.60 percent	6,238	11.8 percent
2	9,090	16.67 percent	8,125	15.4 percent
3	9,114	16.71 percent	10,020	18.9 percent
4	9,027	16.55 percent	12,158	23.0 percent

Table 5-8. Song selection distribution (continued)

	Random selection		Based on rating	
5	9,146	16.77 percent	14,293	27.0 percent
Total	54,537	100 percent	52,886	100 percent

The play counts in the random trial were very close to each other, as can be expected with a random selection. For the trial based on song ratings (or *rating biased selection*), the preference algorithm appears to be *linear* from 12 percent to 27 percent for the rated songs. Moving from the five-star rating downward, the linear preference declines around 4 percent with each step down in rating, but the drop doubles from one-star to unrated, with a fall of 8 percent. While one star might seem like the lowest rating, *no rating* proved the black sheep of the lot.

> Your iPod assumes that if you haven't provided a rating for a song, you must want to hear it even less frequently than those songs to which you have assigned your lowest rating. This is a bit like choosing a movie with bad reviews over a movie that hasn't been reviewed.

Figure 5-2 shows the effects of different song selection options. You can judge the randomness of the true *random selection* option by seeing if those "Random" bars in the figure all seem the same height. The linear nature of the "Rating Biased" bars can be judged by imagining whether there are equal jumps in height as one moves from a rating of 1 to a rating of 5.

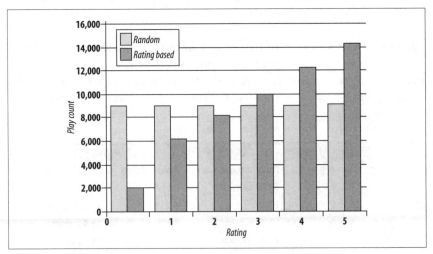

Figure 5-2. Patterns of song selection

Calculating the Statistics of the Selection Process

Changing the number of songs within each rating changes the probabilities for each song's selection. With multiple songs of each rating, the chance of a song with rating r coming up next in the ratings-biased Party Shuffle can be calculated using this expression:

$$\frac{x_r P_r}{(x_0 P_0) + (x_1 P_1) + (x_2 P_2) + (x_3 P_3) + (x_4 P_4) + (x_5 P_5)}$$

Subscripts in this expression indicate the song rating. The chance of a song being chosen is based on x (number of songs with each rating) and P (the proportional weight assigned by the iTunes algorithm for each rating).

With iTunes's preference probabilities for each rating determined from the weekend-long sampling run, here's the resulting expression:

$$\frac{x_r P_r}{0.0388x_0 + 0.1180x_1 + 0.1536x_2 + 0.1895x_3 + 0.2299x_4 + 0.2703x_5}$$

Although the higher-rated songs are given preference, you will not definitively hear more five-star rated songs than all other songs. Let's assume most people follow a normal distribution for their ratings [Hack #23], with the three-star rating being the most common. Table 5-9 displays a hypothetical iTunes library with this bell-shaped curve for the rating song count.

Table 5-9. Typical song rating distribution

Song rating	Number of songs
None	72
1	321
2	1,527
3	1,812
4	507
5	95

If I run these hypothetical numbers through our frequency equations, I get a distribution that looks like Figure 5-3.

As you can see in Figure 5-3, the chance of a song with a particular rating coming up next in the playlist is greatly determined by the song count within the rating. The iTunes preference for higher-rated songs and dislike for lower-rated songs only slightly raises or lowers the probability determined first from the song count.

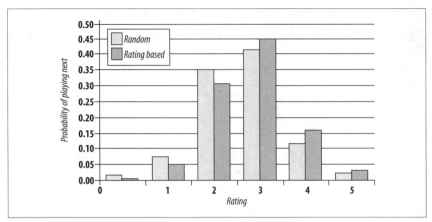

Figure 5-3. Probability distribution of song selection

These chances of hearing a song with a certain rating can be applied to find the chances of hearing a particular song. If we remove the song count from the numerator in the song selection expression, we can calculate the chance of a certain specific song, not just the rating, coming up next:

$$\frac{x_r P_r}{(x_0 P_0) + (x_1 P_1) + (x_2 P_2) + (x_3 P_3) + (x_4 P_4) + (x_5 P_5)}$$

Explaining Statistical Surprises

About a month after running these tests, I noticed my iTunes Party Shuffle at work played the same song two times in a row. This was the first time I had noticed a consecutive repeat, and I checked the playlist. Not only did I find Nirvana's "Territorial Pissings" listed twice in a row, but A.F.I.'s "Death of Seasons" was listed twice in a row three tracks later.

I use the "Play higher rated songs more often" option, but these were each middle-of-the-road 3-star songs, and my song library has nearly 4,000 songs. The odds might seem outrageous at first, but you have to realize just how many songs you hear throughout a workday. If I average 10 hours at work each day and average a 3½-minute song duration, odds say I should hear a consecutive repeat in less than a month.

Many claim to still see patterns as iTunes rambles through their music collection, but the majority of these patterns are simply multiple songs from the same artist. Think of it this way: if you have 2,000 songs and 40 of them are from the same artist, there is always about a 2 percent chance of hearing them next with random play. Right after one of their songs finishes, odds show a 50 percent chance a song by the same artist will play again within the next 35 songs and a 64 percent chance they will be played again within the next 50 songs. This can be calculated following this equation:

$$P(n) = 1 - \left(\frac{x_{\text{total}} - x_{\text{artist}}}{x_{\text{total}}} \right)^n$$

As we have seen in other hacks, a low likelihood event (such as our 2 percent chance of repeating an artist) becomes a highly likely event after just a few opportunities [Hack #46].

It's simply the mind's tendency to find a pattern that makes you think iTunes has a preference.

See Also

Additional technical information about iPods and shuffling can be found at these sources:

- Levy, Steven. "Does Your iPod Play Favorites." January 31, 2005. *http://msnbc.msn.com/id/6854309/site/newsweek/*.
- Hofferth, Jerrod. "Using Party Shuffle in iTunes." August 22, 2004. *http://ipodlounge.com/index.php/articles/comments/using-party-shuffle-in-itunes/*.

—Brian Hansen

H A C K
#55
Predict the Game Winners
The information provided by correlations allows for predicting any outcome, especially sports. With multiple regression techniques and a little software, you can guess the winner before the game is played. The trick is picking the right predictors.

The conventional use of correlations [Hack #11] is to find out how much two variables share in common—or, more technically, how much *variance* is shared between the two variables.

Shared variance is a mathematical term to describe the amount of redundant information reflected in two variables. When lots of variance is shared, prediction is easy and accurate because knowledge of one variable leads to knowledge about a second. Shared variance is estimated by squaring the correlation.

But our everyday world consists of way more than only one variable predicting another. In fact, in most cases there are several or multiple variables that predict a particular outcome. Here we are not dealing with the prediction of just one variable from another, but the prediction of one variable from several. This tool is called *multiple regression* (because there is more than one predictor variable).

Serious sports gamblers, bookies, and casino operators are familiar with multiple regression, or at least they should be. So much information is available about sports teams that there are almost certainly all sorts of variables that, in the right combinations, can fairly accurately predict which team will win.

Betting on professional football is one of the most common of all gambling practices (or so I have been told). This hack shows how to gather data and use multiple regression to predict the winner of any football match up. This example involves predicting who will win the Super Bowl, the National Football League's championship game.

Choosing Predictor Variables

The first step is to build your *model* (the predictors and their weights that you will use to make your prediction). For football, there are dozens of statistics kept and available about teams' past performances and player characteristics. Some make sense as predictors of future performance (e.g., past performance), while others do not (e.g., cuteness of the mascot). The chance to win money, though, is a powerful motivator, so I would take the time and effort to collect just about every statistic I could find about every team and every game. The key is to find variables that on their own correlate pretty well with winning the Super Bowl.

Let's pretend that you have done your research and found six variables that correlate with whether a team wins or loses. Some make sense; some do not. You are interested in getting the most accurate real-life prediction you can get, so you are willing to include the kitchen sink if it will make a difference. To be clear, you took each year that a team was in a Super Bowl and then gathered data for that team from that year.

Imagine you've found that the following variables are of interest and might be useful in predicting this outcome based on previous years' performance and the characteristics of 30 teams. The variables you'll be using in your model begin with the outcome of interest—namely, did the team win the Super Bowl during the year that the data is gathered from (Yes = 1, No = 2)?

The following variables were found to correlate with the outcome:

- Number of easy wins during the season (won by more than nine points)
- Average attendance during the season
- Average number of hot dogs sold per game

- Average temperature of team's Gatorade
- Average weight of defensive linemen

When you do this analysis with real data, you'll likely find a different mix of potential predictors.

Entering the Data into a Spreadsheet

Social scientists often use statistical software such as SPSS or SAS, but for this example, I used an Excel worksheet and Excel's very cool Data Analysis Toolpack (and the Regression Tool). I entered some made-up but realistic data into the spreadsheet shown in Table 5-10.

> What? You thought I was going to show you a real secret formula for predicting the outcomes of football games? I'm only showing you how to make your own. I'll keep mine to myself, thank you very much!

Table 5-10. Super Bowl predictors

Team	Won Super Bowl?	Easy wins	Attendance	Hot dogs	Gatorade	Weight
A	1	11	56,533	4,798	56	276
B	2	9	44,543	5715	76	311
C	1	8	45,543	9,753	45	315
D	1	6	45,768	8,020	46	311
E	1	8	76,786	5,395	56	256
F	1	11	56,533	1,054	67	277
G	2	9	56,554	750	76	256
H	2	12	44,675	6,576	77	254
I	2	11	56,667	9,187	77	287
J	2	10	65,545	4,533	87	301
K	2	12	78,756	1,963	86	243

Table 5-10 shows some of the 30 rows of fictional data I collected, representing 30 examples I used in my statistical analysis. The more rows of data, the more instances you can get and the more accurate your eventual predictions will be.

Building a Regression Equation

You might remember from your high school days that the formula for a simple straight line looks something like this:

$$Y' = bX + a$$

This equation is made up of the following variables:

Y' Predicted score on variable Y

b The slope of the line

X The score of a single predictor

a The intercept (where the straight line crosses the Y or vertical axis)

So, for example, if you wanted to predict human height from weight and had a bunch of data to create such a formula after plugging in the various values, you might get something that looks like this:

$$Y = 35X + 20.3$$

This means that if your weight (the X variable) is 125 pounds, the prediction is that you will be about 64 inches tall, or about 5 feet 3 inches.

But when we have more than one predictor variable, things get more interesting and more fun. There is a longer series of predictors (many Xs) and weights (many bs).

I ran a multiple regression analysis using this data in SPSS, a statistical software program, but you can get much of the same information using Excel (see the "Getting Regression Info in Excel" sidebar).

Getting Regression Info in Excel

There are two ways to get statistical regression info using Excel. First, you can use the SLOPE and INTERCEPT functions, which you can find on the Insert → Function menu. Select the function and enter the argument (the cells where the data is located), and Excel returns these values, allowing you to plug in known values and predict others. This method works best when there is just one predictor.

You can also make use of the Regression option in the Data Analysis ToolPak, an Excel add-on (which you might have to install). Using this option on the Tools menu, you can test the significance of the regression coefficient using an F test, a statistical test similar to a t test [Hack #17].

The results (a.k.a. the *output*) are shown in Tables 5-11 and 5-12. Let's see which of the variables best assist us in predicting whether a team will win the Super Bowl.

Table 5-11. Regression statistics

Multiple R	R square	Observations
0.8483	0.7196	30

Table 5-12. Regression equation

Variable	Coefficients	T stat	P-value
Intercept	−0.784	−1.010	0.323
Easy wins	0.119	4.274	0.000
Attendance	0.000	−0.822	0.416
Hot dogs sold	0.000	1.043	0.308
Gatorade	0.013	2.457	0.022
Weight	0.001	0.580	0.567

Table 5-12 shows a coefficient (a weight) for each of the five variables that were entered into the equation to test how well each one predicts Super Bowl wins. For example, the coefficient associated with "Easy wins" is .119.

If we combine all of these into one big equation for predicting Super Bowl outcomes, here's the model we get:

$$Y' = bX_1 + bX_2 + bX_3 + bX_4 + bX_5 + a$$

So, for each of the predictors (variables X1 through X5), there is specific *weight* (the bs in the formula or the coefficients in the results).

Now, the same formula in English:

b^*Wins + b^*Average Attendance + b^*Hot Dogs + b^*Temp + b^*Weight + a

And using the numbers from the output shown in Table 5-12, here's the real live regression equation:

$$Y' = .119X_1 + .000X_2 + .000X_3 + .013X_4 + .001X_5 + a$$

Interpreting and Applying the Regression Equation

Imagine using this equation with all the rows of data you entered into your spreadsheet. There would be a pretty high correlation between the actual Super Bowl outcomes and the predicted outcome. I know this because of the "Multiple R" part of the output shown in Table 5-11, which shows a pretty high correlation. 0.84 is close to 1, which is the highest correlation you could get.

The "R square" of .72 is the proportion of *shared variance* that we talked about earlier in this hack.

What does this mean? The combination of these predictor variables is a pretty effective way to judge whether a team will win the Super Bowl. Foolproof? Of course not, since the combination of these variables does not perfectly predict the outcome, but it does a pretty solid job.

So, let's say that this year's Denver Cannonballs has the data points shown in Table 5-13.

Table 5-13. Data for Denver Cannonballs

Variable	Value
Easy wins	13
Attendance	35,678
Hot dogs	4,567
Gatorade	65
Weight	267

Plugging this data into the equation shown earlier, here's what we get for a predictor of Y:

$$Y'$$
$$= .119(13) + .000(35,678) + .000(4,567)$$
$$.013(65) + .001(267) - .784$$

The final value for Y' is 1.875, a bit closer to 2 (meaning they are not predicted to win) than to 1 (meaning they are predicted to win).

What's the key to a good set of predictors?

- All the predictors should be independent of each other (if at all possible) since you want them to make a unique contribution to the understanding of what you are predicting.

- Each of the predictors should be as highly related as possible to the outcome that you are predicting.

Improving Your Regression Equation

A careful examination of the equation produced in this hack indicates that the bulk of the predictive power comes from just two variables: the number of easy victories and the temperature of the team's Gatorade. Also, many of the predictors have zero weights, which means you don't need them at all. You could remove these unhelpful variables (attendance and hot dogs sold) to streamline your formula. In fact, collecting data on easy wins and Gatorade temperature alone is enough to make fairly accurate predictions in our example.

—Neil Salkind

Predict the Outcome of a Baseball Game

HACK #56

Turn your radio on in the middle of a baseball game for five seconds and then turn it off. Without hearing the score, you'll be able to name the winner, and you'll be right more than half of the time.

Look, I'm a busy guy. I'm always looking for a way to save time on the less important things in life, such as following my local baseball team, so I'll have more time to spend on the important things in life—friends, family, debating the logic of the Holms' sequential Bonferroni procedure as the appropriate follow-up method to analysis of variance, and so on. A case in point happened just the other day. Wanting to know whether the Kansas City Royals would win a baseball game that was in progress, I hardly had time to wait until the game was over. I wanted to know right *now*!

> Much like Veruca Salt and her interest in owning one of Willy Wonka's Oompa-Loompas "now!", I don't have much patience.

Like a bolt from the blue, I realized that I could turn on my car radio for just a few seconds and have enough information to guess the outcome of the game. And I could do that without hearing the score or who was on base.

How It Works

During the first couple hours of a baseball game, turn on the radio broadcast of that game. Listen just long enough to identify the team that is at bat. That team has a greater than 50 percent chance of winning that game.

Why It Works

Baseball is a game where the longer you are on offense, the more points you can score. As more batters come to bat in a single inning, the chances of moving runners along the base paths and across home plate increases. Another way to look at it is to imagine the end of an inning that was huge for one team. If a team scored a lot of runs, they had to have used considerably more than the minimum of three batters in that inning and, consequently, been at bat a proportionately longer length of time than the other team. Over the course of a game, the team that is at bat longest is more likely to score more (or have more productive innings).

Sampling theory [Hack #19] suggests that a sample is most likely to capture the most common elements of a population. Our population here is all the moments during a game that we could listen to. The most common characteristic in the population (in terms of who is at bat) belongs to the team that is at bat the most.

Figure 5-4 suggests a possible distribution of at-bat time for a regulation nine-inning game. In this example, the winning team was on offense for 58 percent of the time. In retrospect, a random tuning in to the broadcast had a 58 percent chance of finding the winning team at bat.

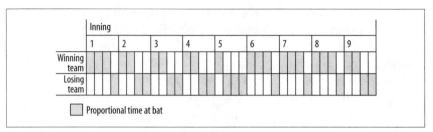

Figure 5-4. Time at bat for winning and losing teams

The accuracy of prediction should be above 50 percent over the long run of baseball broadcasts, but it won't be really, really accurate. This is because the relationship between time at bat and scoring a victory is not a perfect correlation [Hack #11]. Players can score quickly—hit a homerun on their first pitch, for example—or they can take their time getting many hits but strand many runners and never score.

Overall, the correlation between the two variables should be positive, however. Even the perhaps unimpressive 58 percent accuracy in my imagined data in Figure 5-4 means that you will be right 16 percent more often than a blind guess. With such an advantage at the blackjack tables, you would be a millionaire in a week.

Proving It Works

To test the accuracy of my claim, you can use the data that appears in your daily newspaper. While most box scores do not include information about total *time-at-bat* for each team, there is a variable that provides almost the same information. There will almost certainly be a "total at-bats" reported. While this statistic is not the same as time spent at bat, it should correlate pretty highly. Each day, this information is provided for more than a dozen games, and just a few days' worth of data should be enough to test my theory. Gather the total at-bats for each team, including which team won the game.

> Real-life researchers often don't have access to the variable they would really like to know about, and us using *number of at-bats* instead of *time at bat* is good example of this. Instead, we must settle for the next best thing available. Scientists call these substitutes *proxy* variables or *surrogate* variables.

My hypothesis is that the team with the most at-bats should win the game more than 50 percent of the time. Out of curiosity, I tested this hypothesis myself. I used the Chicago Cubs as an example, because their stats were readily available on the Web. I arbitrarily chose 2003 and the Cubs' first 25 games. An analysis of these games found that the team with the most at-bats won 56 percent of the time. If I had eliminated the three situations where there were ties in at-bats, I could have predicted with 63 percent accuracy.

While the team with the fewest at-bats sometimes did win the Chicago Cubs games, the larger the discrepancy between at-bats, the more likely the team with the most at-bats was to win the game. When the most-at-bats teams won, they averaged 4.14 more at-bats than the loser. When the least-at-bats teams won, they averaged only 2.88 at-bats less than the loser.

Other Places It Works

Some people have suggested that in the case of my team, the Kansas City Royals, if I want to be right more than half the time, I should always predict a loss. Yes, yes, very funny.

Where It Doesn't Work

The accuracy of this method should be low if you turn on the radio in the ninth inning, which is why I suggest you try it during the first couple hours of the game. Under the rules of baseball, if the home team is leading after the top of the ninth inning, they never come to bat. They win. Game over. As home teams win more often than visiting teams, this means that often the winning team never comes to bat at all in the ninth inning.

This presents an interesting variation of this prediction method that applies only to the ninth inning. Turn on the game in the ninth inning; if your team is batting, things don't look so good. The data presented for the Chicago Cubs that found the winning team occasionally having fewer at-bats than their opponent can be partly explained by the fact that the winning team sometimes bats in only eight innings.

This method doesn't work for all sports. In basketball, for example, time of possession wouldn't be expected to positively correlate with points scored and, in the case of high-energy, fast-scoring teams, might even negatively correlate. In football, on the other hand, time of position is considered a key indicator of quality performance and usually correlates with a win.

HACK #57   Plot Histograms in Excel

Use Microsoft Excel to plot data distributions so that you can have a better understanding of statistics.

There is some truth to the cliché "a picture is worth a thousand words." A picture is often the best way to understand 1,000 numbers. People are visually oriented. We're good at looking at a picture and observing different characteristics; we're bad at looking at a list of 1,000 numbers.

One of the most powerful tools available for understanding data is the *histogram*, a picture of the distribution of values. Here is the idea of a histogram. Suppose you have a lot of data—say, the batting averages for all 6,032 baseball players between 1955 and 2004 who averaged 3.1 or more plate appearances per game. Let's also assume you want to know how these values are distributed. What are the lowest and highest values? Are there more low values than high values? Were batting averages totally random numbers between 0 and .400, or was there some pattern?

Batting average can take many different values. Between 1955 and 2004, 6,032 players had qualifying batting averages, and there were 1,229 unique values for batting average. You can plot the number of players with each unique batting average (though I can't imagine what this graph would look like). But we don't really care about each unique value; for example, the fact that 13 players had a batting average of .2862 is not that interesting. Instead, we might want to know the number of players with very similar batting averages—say, between .285 and .290.

Let's think of each range as a bucket. Every player-season goes into a bucket. For example, in 1959, Hank Aaron had a .354 average, so we'll put that season in the .350–.355 bucket. So, here's our plan: we'll put each player-season into a bucket, count the number of player-seasons in each bucket, and draw a graph showing (in ascending order) the number of players in each bucket. This single diagram is a histogram.

The Code

In this example, I wanted to look at the distribution of batting average. I used a table containing the total batting statistics for each player in each year (and the list of all teams for which each player played), and I called the table b_and_t. I selected only batters with enough plate appearances to qualify for a league title, and only those players who played between 1955 and 2004:

```
SELECT b.playerID, M.nameLast, M.nameFirst, b.yearID, b.teamG,
b.teamIDs, b.AB, b.H,
b.H/b.AB AS AVG,
b.AB + b.BB + b.HBP + b.SF as PA
FROM b_and_t b inner join Master M
on b.playerID=m.playerID
WHERE yearID > 1954
AND  b.AB + b.BB + b.HBP + b.SF > b.teamG * 3.1;
```

After running this query, I saved the results to an Excel file named *batting_averages.xls*.

One way to draw histograms in Excel is to use the Analysis ToolPak add-in. You can add this by selecting Add-Ins... from the Tools menu, and then selecting Analysis ToolPak. This adds a new menu item to the Tools menu, called Data Analysis, which introduces several new functions, including a Histogram function. But I find this interface confusing and inflexible, so I do something else.

Here is my method for creating a histogram:

1. In the data worksheet, create a new column called Range.

2. In the first cell of this column, use a function to round the value for which you would like to plot the distribution. The simple way to do this is to use the Significant Figures option of the ROUND function. In my worksheet, column I contained the value for which I wanted to calculate the distribution (batting average), so I could use a formula such as ROUND(I2,2) to round to the nearest .010. Personally, I find a bucket size of .005 to be more descriptive, so I use a trick. You can multiply a value inside the ROUND function and then divide outside the function to get buckets of almost any size. Inside the ROUND function, I multiply by the reciprocal of the bucket size—in this case, 1 / .005 = 200. Outside the function, I multiply by the bucket size. In my worksheet, column I contained the average values. So, I used ROUND(I2 * 200,0) / 200 as my formula. Copy and paste this formula into every row of the worksheet. (You can double-click the bottom-right corner of the cell to do this quickly.)

3. Now, we're ready to count the number of players in each bucket. Select all of the data in the worksheet, including the new Range column. From the Data menu, select Pivot Table and Pivot Chart Report. Select Pivot Chart Report and click Finish (we'll use all the defaults). We will select

two fields for our pivot table. From the Pivot Table Field List palette, select Range. Drag-and-drop this onto the Drop Row Fields Here part of the pivot table. Next, drag-and-drop "playerID" onto the Drop Data Item Here part of the pivot table. By default, Excel will count the number of player IDs in the underlying data that match each range value. The pivot table is now showing the number of items in each bucket. You should see a (very ugly) graph with the number of players in each bucket.

4. Clean up the graph. (I like to erase the background fill and lines and change the width of the columns.) Figure 5-5 shows an example of a cleaned-up graph.

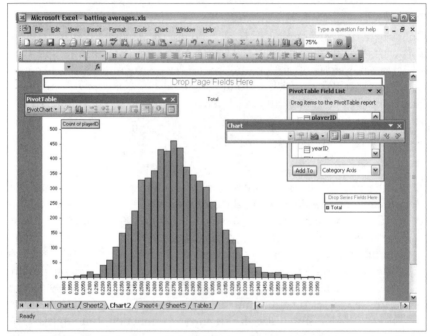

Figure 5-5. Histogram from a pivot chart report

Looking at the histogram, we see that the distribution looks similar to a bell curve; it skews toward the right and is centered at around .275.

Hacking the Hack

One of the nice things about calculating bins with formulas is that you can easily change the formula for binning. Here are a few suggestions for other formulas:

ROUNDDOWN(*<value>, <significance>*) *and* ROUNDUP(*<value>, <significance>*)

This ROUNDDOWN function rounds down to the nearest significant figure. For example, ROUNDDOWN(3.59,0) equals 3, and ROUNDDOWN(3.59,1) equals 3.5. Similarly, ROUNDUP rounds up to the nearest significant figure. ROUNDUP(3.59, 0) equals 4, and ROUNDUP(3.59,1) equals 3.6.

LOG(*<value>, <base>*)

Sometimes it's useful to plot a value on a logarithmic scale, and to use logarithmic-size bins. You can combine LOG functions with ROUND functions to create variable-size bins.

CONCATENATE(...)

The CONCATENATE function doesn't compute numbers, it puts text together. If you want to explicitly list ranges (such as 3.500–3.599), you can use the CONCATENATE function to create these; for example, CONCATENATE(ROUNDDOWN(3.59,1)," to ",ROUNDUP(3.59,1)-0.01) returns 3.5 to 3.59.

If you want to take this to the next level, you can replace the bin size with a named value. (For example, name cell A1 *bin_size*.) This makes it easy to change the bin size dynamically and experiment with different numbers of bins.

—*Joseph Adler*

Go for Two

HACK #58

In football, when is the two-point conversion attempt the right choice? Regardless of which "chart" you're using, the problem gets even more complicated when statisticians enter the debate.

A few years back, I was enjoying watching my local professional football team as they were losing a close game. I wasn't entertained by my team's dismal performance as much as I was delighted by my team's befuddled coach as he attempted to read and understand a *two-point conversion chart*.

In football, after a touchdown is scored (the touchdown itself is worth six points), the scoring team has two options for scoring an "extra point" or two. Usually, the team chooses to kick a single extra point through the uprights (like a short-distance field goal), but they might also choose to "go for two" points (known as the *two-point conversion*), which involves the offense rushing or passing for another trip into the end zone.

At the time, as was later "confirmed" by sportswriters, it was clear that he wasn't sure how to read the chart. Specifically, when interpreting the column on the chart that listed how many points behind or ahead a team was, he thought this meant how many points ahead or behind a team *would be* if they made the point-after conversion.

As I mused about how an NFL head coach might never have learned to read such a chart, I began to wonder who produced this "chart" and what principles it was based on. Later, as I searched for the "official chart," I found two "official" charts, and they didn't always agree.

More recently, I ran across a chart based on a statistical analysis of the probability of possible outcomes and on the amount of time remaining (as indicated by the number of possessions remaining). This chart didn't agree with either of the earlier charts I discovered.

This hack is for you, Coach. It examines from a statistical perspective when to go for two points and when to settle for one.

Traditional Two-Point Conversion Charts

When you see a coach on TV holding a plastic laminated card and studying it before deciding whether to go for two, sportscasters like to refer to the card as *the chart*, though, as mentioned in the previous section, there's more than one chart in use. The slight differences might be due to the fact that one is identified as being used in the NFL and the other is identified as a classic set of standard decisions used in college football.

The differences might also be based on the fact that the college chart was produced for a certain team that may have had a more aggressive or confident style. The college chart seems to play for a victory, not a tie. Though college ball now has overtime rules, they are a fairly recent development, whereas the pros have had overtime for a while.

The NFL chart is provided on Norm Hitzges' web site (Norm is a broadcaster in Dallas and an all-around sports guru) at *http://www.normhitzges. com/thechart.htm*. The college chart (found at *http://www.NFL.com/fans/ twopointconv.html*) is identified as the one used in the 1970s and developed at the University of California, Los Angeles (UCLA). Table 5-14 provides the suggested decisions from both charts and is condensed a bit.

Table 5-14. Classic decision making for two-point attempts

	Points behind or ahead												
	0	1	2	3	4	5	6	7	8	9	10	11	12
Behind (NFL)	1	1	2	1	1	2	1	1	1	1	2	1	1
Behind (College)		2	2	1		2	1	1	1	2	1	2	2
	0	1	2	3	4	5	6	7	8	9	10	11	12
Ahead (NFL)	1	2	1	1	2	2	1	1	1	1	1	2	2
Ahead (College)		2	1	1	2	2	1	1	1	1	1	1	2

The UCLA chart does not provide suggestions for when the score is tied or when your team is behind by four points. The NFL chart, on the other hand, is full of advice for all occasions. As discussed, the primary difference seems to be whether you're willing to play for the tie or not. UCLA clearly did not wish to play for the tie, while the NFL chart has no such hesitancy.

Modern Super-Scientific Chart

In the real world, a set of statistical probabilities controls the outcome of a sporting event, and the decision about whether to go for two or take the extra point should be based on more information than just the score and whether your team is winning or losing. In actual game situations, smart coaches take the following additional factors into account:

- The likelihood that their field goal kicker will make the field goal
- The likelihood that their team will score on a given two-point conversion play
- The current health, attitude, and skill of their players
- How many more possessions their team will receive

Past statistics show that the average NFL football team makes about 98 percent of its extra points and about 40 percent of its two-point attempts. Coaches must use their experience and intuition to gauge their players' current ability level, and a chart isn't much help on that score.

As for possessions left, however, this is exactly the type of information that decision systems based on probability need to take into account. Based on a process of working backward from the ending of a hypothetical football game that takes the probability of success on either option (98 percent for one-point plays and 40 percent for two-point plays) into account, statisticians have produced a chart based on not only on the current score, but also on the total number of possessions remaining for both teams.

In a 2000 issue of *Chance* magazine (Vol. 13, No. 3), Harold Sackrowitz presented the results of such an analysis using a process called *dynamic programming*. Table 5-15 shows a portion of Dr. Sackrowitz's chart.

Table 5-15. Modern decision making for two-point attempts

Possessions remaining		Points behind or ahead												
		0	1	2	3	4	5	6	7	8	9	10	11	12
1	Behind	1	1	2										
	Ahead	1	2	1										
2	Behind	1	1	2	1	1	2							
	Ahead	1	2	1	1	1	2							
3	Behind	1	1	2	1	1	2	1	1	2				
	Ahead	1	2	1	1	1	2	1	1	1				
4	Behind	1	1	2	1	1	2	1	1	2	1	2		
	Ahead	1	2	2	1	1	2	1	1	1	2	1		
5	Behind	1	1	2	1	1	2	1	1	2	1	2	1	
	Ahead	1	2	1	1	1	2	1	1	1	2	1	1	
6	Behind	1	1	2	1	1	2	1	1	2	1	2	1	2
	Ahead	1	2	2	1	1	2	1	1	1	2	1	1	2

This two-point conversion chart is based on the branching possibilities starting at different points in the game and assuming basic probabilities of success for either an extra point or a two-point conversion. An average NFL quarter sees six possessions in total, so think of this chart as being most useful in the fourth quarter. Sackrowitz also assumes a 50 percent chance for overtime victories.

How It Works

The calculations for Table 5-15 work something like this simple example:

1. Imagine you are down by one point without much chance of getting the ball again.

2. You have a 98 percent chance of making an extra point kick and a 50 percent chance of winning in overtime. Going for the extra point results in a victory 49 percent of the time (.98 x .50 = .49).

3. You have a 40 percent chance of converting a two-point play, so going for two points results in a victory 40 percent of the time. Failure ends the game, and success wins the game.

4. 49 percent is better than 40 percent, so you should elect to go for the extra point. Notice that if you believe your team's chances of converting the two-point play are better than 49 percent, you should go for it. Calculations like these, but over a longer series of possessions, result in the decision tree reflected in Table 5-15.

Which chart should you use the next time you find yourself coaching in a crucial football game with a key decision to make? That's up to you, but just remember that befuddled football coach I watched on TV a few years ago. Not only was he replaced the next year by Dick Vermeil, considered one of the brighter football coaches around, but it was Vermeil who helped develop the UCLA two-point conversion chart shown in Table 5-14. Now you know the *rest* of the story!

HACK #59 Rank with the Best of Them

There are many ways to use data to make judgments about who is best in any sport. All the intuitive ways to compare performance in individual sports have validity concerns, however.

My friends and I are a competitive lot. Our arena of combat, most recently, has been poker. On a regular basis, my friends and I gather at my home and take part in a Texas Hold 'Em poker tournament. It's an informal affair, but we all take it very seriously. The way our poker tournaments work, everyone starts with the same amount of chips, and when they are gone they are gone. There is a first one out, a last one out, and everything in between. So, for example, if seven people play, someone comes in first, second, third, fourth, fifth, sixth, and seventh.

We all think of ourselves as pretty good and, being competitive, we have longed for an objective method of comparing performance across tournaments. As one of the statisticians in the group, I took it upon myself to devise various ways of producing some sort of objective index that would allow all participants to compare their performance with each other to decide once and for all who is the best player and who is only lucky now and again. This is the story of my quest and the statistical solutions I chose. Not to give the ending away, but I learned that there is no single best solution.

How to Rank Fairly

This business of how to identify the best is a common problem for competitive organizations such as sports leagues and associations. The problem is how to summarize performance across a variety of categories, venues, and occasions.

There are three methods commonly used in the world of sports to make determinations about who is the "best." All of the approaches make some intuitive sense, though each method has its own specific advantages and disadvantages.

First, let's take a look at the nature of the data I had to analyze. Your data will likely be similar, whether you run your weekly home Monopoly game or you run the Professional Golf Association. Though poker is not a sport, any organized competitive endeavor provides data for rankings. Table 5-16 shows the results from eight tournaments in my own summer poker league.

Table 5-16. Summer poker league data

	Paul	Lisa	Billy	BJ	Mark	Bruce	Cathy	Tim	David
5/14	6	5	4	3	2	1			
5/21	3	6	4	5	7	2	1		
5/28			5	4	1	3	2		
6/4			4	6	3	7	2	5	1
6/11			4	5	6	1	2	3	
6/18			5	4	2	3	1		
6/25			1	4	3	5	2		
7/2			1	5	4	3	2		

You can see that nine players took part in at least one tournament, but no event had participation from all players. If a person received no points on a given night, it was because she didn't play. This is commonly the case in sports such as golf and tennis as well.

On two occasions, seven people played, but on other occasions, as few as five sat down together. Four people have played in all eight tournaments. (These are the hard-core players who have to admit that they have a bit of a problem recognizing what is important in life.) One player, David, played in only one tournament.

The points under each player's name indicate the order in which they went out. If there are six players and you go out first, you get one point for taking last place. If you are the winner among six players, you get six points for taking first.

Notice a couple of things about this point system. First, you get at least a point just for showing up. Second, you get more points for winning a tournament with more players.

How, then, to rank players in the poker league? Here are three common solutions, all of which work to some extent.

Total points. The first thought that came to mind in my situation was to simply add up the points across tournaments and rank players based on their total points. This is the approach taken when celebrities are ranked by income or bank robbers are ranked by their number of crimes. Just participating a lot moves you up in these rankings. To be golfer of the year, you have to have played in many events, in addition to performing OK in them.

Mean performance. A second method is to average the points by dividing the total points by the number of tournaments in which a player participated. The beauty of producing an average is that you get a number that represents a typical level of performance. This is ideal for measuring something elusive, such as talent. Your average performance at poker (or anything else) should be the best single indicator of ability.

Total wins. A third method, the simplest and most commonly used in team sports, is to count victories. The player who wins most often is the best player. This method works well for tournament-style poker (the kind we play) and any events in which there is one competitor who is the clear winner.

Comparing the Three Methods

Though each ranking approach has some clear advantages and does the job adequately, Table 5-17 shows the values for each player under all three ranking systems.

Table 5-17. Summarizing poker performance

	Paul	Lisa	Billy	BJ	Mark	Bruce	Cathy	Tim	David
Points	9	11	28	36	28	25	12	8	1
Mean	4.5	5.5	3.5	4.5	3.5	3.13	1.71	4.0	1.0
Wins	1	1	2	1	2	2	0	0	0

All three scoring systems make sense. But the question about who is the best has a different answer under each of the three systems! This is certainly a frustrating finding for a poker scientist like me. Because one could defend any of the three methods as the "best" way to rank, it is a bit of a paradox that each method produces a different "best" poker player. Table 5-18 shows how the rankings differ under each scoring method.

Table 5-18. Poker rankings

	Paul	Lisa	Billy	BJ	Mark	Bruce	Cathy	Tim	David
Points	7	6	2.5	1	2.5	4	5	8	9
Mean	2.5	1	5.5	2.5	5.5	7	8	4	9
Wins	4	4	2	4	2	2	6	6	6

Notice how the "best player" is different under each system. BJ is the best under the Points system. Lisa is the best under the Mean system. Three people tie for first under the Wins system, but BJ and Lisa are not among them. The only real agreement across the three methods is that David is ranked as the worst player. (Sorry, David, but numbers don't lie. And sorry about the public ridicule. Maybe I can make it up to you with a free copy of this book?)

> I broke ties when assigning rankings by averaging the ranking among those who were tied. In other words, Billy, Mark, and myself were all tied for the number one ranking under the Wins system, so the ranks of 1, 2, and 3 average to 2, and that was our ranking.

If three different scoring systems result in three different rankings, it is clear they cannot all be equally valid. They cannot all produce scores that truly reflect the variable of interest, which is poker-playing ability defined in the same way. The solution does not involve picking the single best approach. It was not my goal to identify the best system and go with it; my goal was to provide valid information and let others interpret the data how they want.

My solution was to provide all three rankings based on the three scoring methods. That way, players could choose to focus on the ranking results from the method that makes the most sense to them.

The End of the Story

The system that made the most sense to the players in my poker league turned out to be the one that ranked them the highest. Imagine that.

I sleep at night secure in the knowledge that any of the methods is probably acceptable and "accurate." After all, none of the three methods makes the mistake of identifying me as the one best player. That's got to be some sort of validity evidence in and of itself!

Real-life professional sports organizations have dealt with the advantages and disadvantages of each system by creating composite point systems. Some of the tinkering to improve ranking systems in tennis and golf (and tournament poker, too) includes:

- Combining performance data over a long period of time
- Awarding more points for winning more difficult tournaments
- Using *both* the mean performance and total points together to reward excellence *and* frequent participation

It is a bit ironic that these systems that are likely fairer and more accurate are often perceived by the press and fans as overly complex and crazy. Attempts to make the ranking systems more *valid* have resulted, often, in a rejection of the systems by the public as *invalid*.

Estimate Pi by Chance

Statisticians like to think that anything important can be discovered using statistics. That might actually be true, since it turns out that you can use statistics to estimate the value of one of the most important basic values in science: pi.

The ability to calculate pi is one of the routine skills for all budding geniuses. I remember, for example, that dividing 22 by 7 comes pretty close. There are a variety of other ways, some more accurate than others. My favorite method, though, requires the element of chance and a long, lonely sea voyage or other period of enforced solitude. Intrigued? Read on, Gilligan.

Before showing how to estimate the value of pi, I'll begin our discussion by presenting a couple of basic facts from geometry. Don't panic; I don't know much about geometry, so we won't spend a lot of time on this. I'll just cover the basics we need to appreciate the magic of this hack.

Pi

In geometry, key relationships have been found between *pi*, a number that is roughly 3.14159 (symbolized by π), and the way various parts of a circle fit together, as shown in Figure 5-6.

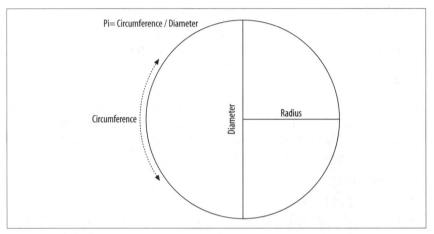

Figure 5-6. Calculating pi

For example, if you take the diameter of a circle and multiply it by pi, you will get the circumference of the circle. If you take the radius of a circle, square it, and multiply that value by pi, you will get the circle's area.

All pretty cool, perhaps, but it is primarily of interest to those who like to play with geometry, not with statistics. But just wait.

Pi and Falling Needles

In the 1700s, Georges-Louis Leclerc presented a half-geometry/half-statistics puzzle to the world. He was the Count of Buffon, or something, so this problem is known as Buffon's Needle Problem. He presented it generally, without specifics, and I summarize it here:

> Imagine a needle lands randomly on a drawing of two parallel horizontal lines. The lines are further apart than the length of the needle. What are the chances that the needle will land in such a way that it touches one of the lines?

This is one of those problems that seem impossible to solve the first time you hear it, but it *is* solvable. There's no need to spend any time calculating the solution here, though I certainly *could* do it, I assure you. Really, I could. Really. The solution has to do with geometry, and it takes into account two key components of information. The keys to any given random landing position are:

- Where the center of the needle is in terms of distance from the closest line
- The angle of the needle in relationship to the perpendicular of the closest line

Defining the random position of the needle with these two bits of information allows for some general observations that help to simplify the problem:

- If the center of the needle is exactly on one of the lines, then the needle will *always* touch that line, regardless of its angle.
- If the center of the needle is close enough to a line, within half the needle's length, then the needle will *sometimes* touch a line. The angle of the needle determines whether the needle touches a line.
- If the center of the needle is further away from a line than half the needle's length, then the needle will *never* touch that line, regardless of its angle.
- The closer to a line, the greater the chances are of a needle touching that line.

All the possible needle locations can be graphed as a curve, illustrating all possible distances from a line and all possible angle-from-perpendiculars of the needle. Trigonometry enters the picture here, and mathematicians have defined such a curve with this equation:

$$\text{Probability} = \frac{(2)(\text{length of the needle})}{(\pi)(\text{distance between the lines})}$$

This is the answer to the problem. Let's try it quickly with some real numbers, just to check Leclerc's work. Imagine a needle three inches long falling randomly on a sewing table with a pattern on the grain such that there are two parallel lines four inches apart. What proportion of the time will the needle touch one of the two lines? Here are the necessary computations:

$$\frac{(2)(\text{length of the needle})}{(\pi)(\text{distance between the lines})} = \frac{(2)(3)}{(3.1459)(4)} = \frac{6}{12.566} = .477$$

The needle will touch a line about 48 percent of the time.

Already, your gambling juices might be flowing as you envision a large room full of needle-dropping and lines on the floor and such. Go for it; more power to you. This principal is already in play in some carnival games you've probably seen. Ever notice how rarely those ping-pong balls land in those fishbowls or the football gets through that hoop?

Probability and Pi

I promised you that you could use chance to estimate pi, though, not use pi to figure chance. The power of math allows us to move around any element of any equation, and so any element to the right of the equals sign can be moved to the left. We can scramble our probability equation to produce a pi equation like so:

$$Pi = \frac{(2)(\text{length of needle})}{(\text{Probability})(\text{distance between lines})}$$

I'll prove it works by using the same numbers we used when we tested the probability equation. We already know what the right answer for pi is, so let's see if the equation works:

$$\frac{(2)(\text{length of needle})}{(\text{Probability})(\text{distance between lines})} = \frac{(2)(3)}{(.477)(4)} = \frac{6}{1.908} = 3.1447$$

This equation calculates pi as 3.1447, which is pretty darn close to 3.14159. If we had allowed our numbers to go many places past the decimal, we would have had an even more accurate answer.

Estimating Pi Using Probability

In our example, we knew the probability, so we could calculate pi using that information. But what if you didn't know pi and needed to calculate it? What if you were stuck on a desert island or on a long ocean voyage or in bed with a broken leg and had no access to reference works that included a fairly exact value for pi? Further, suppose you needed to calculate the circumference of a circle or the volume of a sphere or any of a number of other values in geometry or finance or physics that make use of the pi value? A nightmare scenario, eh? You could use this formula to calculate pi pretty accurately by just conducting an experiment and collecting data.

Set up an area with two horizontal lines, drop some needles, and keep track. Measure the distance between your lines and the length of your needle, and let the random whims of chance do all the cognitive heavy lifting. Collect a large sample of data from many needle drops to get a probability that is precise to several places past the decimal, perhaps a thousand drops or so. Good luck and keep careful records.

Let's say that you drew two lines that were 8 inches apart and used a knitting needle about 7 inches long. If you used this equipment for a large number of drops, you would likely find that the needle touched a line somewhere between 50 and 60 percent of the time. Let's say it was 55 percent. To use this data to calculate pi, you would apply the math like this:

$$\frac{(2)(\text{length of needle})}{(\text{Probability})(\text{distance between lines})} = \frac{(2)(7)}{(.55)(8)} = \frac{14}{4.4} = 3.18$$

You'll find that 3.18 is pretty close to the ratio of the circumference to the diameter shown in Figure 5-6.

If your eyesight isn't what it used to be, there's no need to use a hard-to-see needle. You can apply the same logic using a pencil falling off your desk, or a marble rolling across the floor into a defined area, or a parachutist landing on a rectangular target. You need two parallel lines that the pencil, marble, or parachutist can have a chance of landing on, and you need to know the length of the object. As long as the outcome is random, anything will work, and a parachutist landing on a haystack is a lot easier to find than a needle somewhere in one.

CHAPTER SIX

Thinking Smart
Hacks 61–75

This chapter concentrates on hacks that help you to think more clearly, cleverly, or creatively. Start out by using the rules of probability and proving yourself smarter than a superhero [Hack #61]. Keep feeling smart by mastering statistical shortcuts [Hack #66] and the ability to detect fraud [Hack #64].

Continue impressing yourself and others by tapping into your skeptical side: demystify amazing coincidences [Hack #62] and hack your way to the truth about weird phenomena [Hack #63]. After disproving (or perhaps proving) the existence of ESP [Hack #68], your friends will be amazed when you read their minds [Hack #67].

Finally, wrap up your self-improvement course by learning to avoid a common illogical trap [Hack #69].

Now that you are so smart, it should be a breeze to notice things around you that others do not. You can master the fine art of the traffic jam [Hack #74], explore your connections to Kevin Bacon and everybody else [Hack #72], and spot bogus election systems [Hack #73] known only to political scientists.

Round out this chapter by expanding your horizons. Try out different exciting professions such as espionage and code breaking [Hack #70], and discover new species [Hack #71] and, perhaps, even life on other planets [Hack #75].

HACK #61 Outsmart Superman

Lightning can strike twice in the same place, but it is very unlikely. The laws of probability allow us to calculate the likelihood of a series of rare occurrences happening all in a row.

Occasionally, we hear stories of highly unlikely events happening more than once to the same person—a forest ranger who has been struck by lightning seven times, for example, or a New Jersey couple winning the lottery twice.

When they appear in the news, these stories often include an interview with the local stats professor, who estimates the odds of such a thing happening.

The math for calculating the total likelihood of a *series* of events is fairly simple. The more difficult part is figuring good estimates for the probability of any single event happening once. Then, you simply multiply the individual probabilities together to get the total likelihood for the whole chain of weird happenings.

Lucky Lois Lane

To show the steps involved in calculating the likelihood of a whole series of events, I've chosen an example from classic literature. This series of rare events is described in the *Lois Lane* comic magazine #56, published by DC Comics in April of 1965. A common pattern in the stories involved Lois having some apparently supernatural experience that was hard to explain but, at the end of the story, turned out to have some simple explanation.

> Lois Lane, now wife (but former girlfriend and number one fan) of the comic book hero Superman, was a very popular character in the line of DC comic books in the 1960s and 1970s. Among sophisticated comic aficionados, *Lois Lane* comics of that era are now enjoyed as examples of particularly strange comic writing. Lois tended to beat the odds almost on a daily basis. Her comics should be required reading in statistics courses.

One example of a strange experience that was then "explained" by Superman at the end of the story involved the application of a statistics hack. Lois is pretending to be telepathic so she can hang around mobster "Long Odds" Larkin and maybe get a scoop for her newspaper.

It works all too well, as she is kidnapped by Larkin and forced to provide him with "telepathic" information so he can commit crimes. Fortunately for Lois, and for the mobster, her blind guesses turn out to be correct and Larkin keeps her alive. Her guesses are so accurate that Lois comes to believe that she actually has psychic powers.

It turns out, according to Superman, who eventually rescues her, that Lois was just lucky! Very lucky. Astoundingly, incomprehensibly lucky. Even though the odds of Lois correctly making the lengthy series of correct predictions and accurate guesses were extremely slim, she just lucked out. Congrats, Lois!

Superman presents what he says are the odds for the fantastic feats that Lois performs, but the author of the story (anonymous) does not provide the calculations. Let's review the random guesses that Lois makes, do our own calculations, and check the Man of Steel's math. For determining the probability of this series of independent events, we will apply the multiplicative rule [Hack #25].

The Guesses

In the story, Lois correctly guesses—totally at random, mind you—the following:

1. Which of five duplicate armored trucks is actually carrying Metro Bank cash

2. The combination to a safe that holds a large company's payroll funds

3. The unlisted phone number to the richest person in town

4. Under which of 20,000 trees a bank robber's loot is buried

She finally fails, after Superman has rescued her, to guess the number of jellybeans in a jar. As Superman explains to Ms. Lane that she is not psychic, he suggests that the odds of her making these four correct guesses by chance are 326,454,839,047 to 1, or 1 out of 326,454,839,048.

"I see, Superman!" she says. "I was lucky enough to hit that 'one chance'."
"Yes," says Superman, "after all, someone always wins big lotteries, too" (or some such nonsense to that effect). That number calculated by Superman or his Supercomputer certainly is big, which seems right, but I don't think it is close to being correct. My guess is that this outcome is even more miraculous.

The Calculations

Let's work through our own calculations. For guesses 1 and 4, we can figure pretty close to the odds of guessing the answer to that problem independently. For guesses 2 and 3, we'll have to make some assumptions.

Here again are the guesses Lois made and real calculations of the odds for each one, taken by themselves.

> The math involved here is the easy part for statisticians who are asked to produce statements of likelihood for a string of unlikely events. The hard part is determining the starting values, the pieces of the equations. As you see with our attempts to estimate how lucky Lois was, we will have to make some moderately wild, though reasonable, guesses to somehow *know* what the chances of any particular occurrence are. Statisticians can't really know the basic odds much of the time. They tend to focus on theoretical situations where the odds *can* be known, not real-life problems like those of Ms. Lane.

Guess 1. Which of five duplicate armored trucks is actually carrying Metro Bank cash? This is the easiest one. Five possibilities, one correct choice. The chances are 1 out of 5 or 1/5.

Guess 2. Lois guesses the combination to a safe that holds a large company's payroll funds. This is a real puzzler. Not only does Lois guess the five numbers that one should turn the dial to, but she also guesses that there is a sequence of five different numbers that must be used, and the directions that the wheel must be turned.

In the real world, there are a variety of different types of combination locks produced, so it is hard to know for sure what assumptions we should make about this problem. I've done a little research about safe cracking (for the sake of this hack, let's say) and learned a little about combination safes. Usually, there is a total of anywhere from one to eight numbers in a combination sequence. I'd guess that three or five numbers in a sequence is most common. The numbers on a dial can be any range of values, but 0 to 99 is common for larger safes, such as the payroll safe in the story.

So, for starters, let's say that she randomly picks between this safe having a three- or five-number combination. Chances for that guess are 1 out of 2, or 1/2. Say she randomly picks a number from 0 to 99 each time: 1 out of 100, or 1/100, for each number in the sequence. She also has to guess the starting direction. Let's say that most safes, 80 percent, start to the left, and only 20 percent, 1 out of 5, start to the right (which is her guess).

So far, so good. It gets very tricky here, though, because of the combination Lois actually suggests. She predicts "11 right...13 left...5 left...back to 8...forward to 15." This is a very odd combination. First, a combination is usually read in a different order: *left 13*, instead of *13 left*. Second, what can it possibly mean to go left *twice* in a row! Surely you have to change direction of the dial to lock in each number in the sequence. After all, the dial passes over many numbers on its way left every time. How does it know whether to count each number it passes as a part of the combination sequence? I'm going to just pretend that the sequence is misreported slightly by the anonymous author; otherwise, I'd have to pause here in an endless loop of confusion, with my fingers over the keyboard, never able to continue.

Finally, why does Lois start saying "back" and "forward" instead of left and right? This just makes her directions unclear (perhaps to cover herself in case of failure?). Again, I'm going to assume she uses the terms to mean a change in direction, even though *back* probably means *left* and *forward* probably means *right*, which would just complicate things more. A conservative set of probabilities for this guess, then, is $1/2 \times 1/5 \times 1/100 \times 1/100 \times 1/100 \times 1/100 \times 1/100$. That's 1 out of 100,000,000,000.

Guess 3. Lois also guesses the unlisted phone number to the richest person in town. There are a couple of ways to figure this.

First, if Lois were a bit naïve (and, no offense to Lois's fans, but I'm guessing she is), she might set only the parameters that the phone number had to have seven digits and not start with 0. Under these rules, there are 9,000,000 possible phone numbers. This assumes that we start with 10,000,000 possible seven-digit numbers (9,999,999 is the highest seven-digit number, plus add one for the number 0,000,000).

If we can't count any numbers that start with 0, that eliminates the number 0,000,000 and all six-digit or less numbers (there are 999,999 of those). That's an even million possibilities we can eliminate. So, under this scenario, Lois's chance of guessing the number would be 1 out of 9,000,000 or 1/9,000,000. Let's give Lois the benefit of the doubt for a second and imagine that she wouldn't guess her own phone number or other phone numbers she knows by heart. I'd guess there are maybe 10 of those. So, Lois would have 1 out of 8,999,990 to choose from.

A smarter Lois (let's say for the sake of argument) might know the particular exchanges in use in Metropolis, or those likely to be used for unlisted numbers, or for the rich part of town, or whatever. Back in the day, there was a small set of possibilities for the first three digits in a particular area code known as exchanges. A city the size of Metropolis might have fifty or so that were used most commonly, so she might choose from those. Under the "smart Lois" scenario, her odds improve considerably. Now, she might blindly guess out of 500,000 numbers, not 9,000,000. Her chances might have been 1 out of 500,000 or 1/500,000. My rough estimation of Lois's intelligence suggests that this scenario is not the most likely, but she is a reporter for a major metropolitan newspaper, so she may have this knowledge. Let's be charitable and go with it.

Guess 4. Finally, Lois guesses under which of "20,000" trees a bank robber's loot is buried. Like guess 1, this is also fairly easy to calculate. If there really are exactly 20,000 trees in the woods where the loot is buried (and this number is probably an estimate or rounded off), the chance of guessing correctly is 1 out of 20,000 or 1/20,000.

Final Probability

So, the chances of guessing correctly on these four problems in a row, giving Lois all sorts of benefits of the doubt for knowing all sorts of things about safes and telephone numbering systems, is $1/5 \times 1/100,000,000,000 \times 1/500,000 \times 1/20,000$. The chances of this sequence of lucky guesses occurring is, conservatively, 1 out of 5,000,000,000,000,000,000,000—even more remarkable than the already hard to believe 1 out of 326,454,839,048.

"I see, Superman! I was lucky enough to hit that one chance," Lois concludes. Indeed. Of course, the odds were even worse that Superman would propose to Lois someday, and that happened. So, who am I to rain on Mr. and Mrs. Superman's parade?

HACK #62 Demystify Amazing Coincidences

The patterns of probability produce some unusually interesting alignments. Here's how to interpret coincidences that seem unbelievable.

One of the occasional sad duties of statisticians is to take a world full of whimsy, delightful serendipity, and surprises around every corner and turn it into a dull, predictable, uninteresting place. I'm about to do that here, so if you would rather keep wearing rose-colored glasses, put them on now, skip this hack, and pick another one (I suggest more pleasant topics, such as winning Monopoly [Hack #51]).

I choose to be scientific and treat the world as rational and built on consequences that follow chains of cause and effect. My problem—and perhaps yours too, if you think like me—is that when I face anomalies (hard to explain, unexpected things), it is tempting for me to treat the happening as evidence of something mystical, or psychic, or paranormal in some way. Coincidences are a good example. When I witness an incredible coincidence, I am tempted to fall into a comforting pit of nonscientific explanations, such as fate or synchronicity.

Synchronicity is the term that pioneering psychiatrist Carl Jung used for personally meaningful coincidences. He saw them as providing insight into the inner world of the unconscious, but was not above assigning pseudo-mystical explanations to them as well. He was not a statistician.

The solution to my problem—and perhaps yours, if you're still with me—is to think a bit and apply some basic rules of probability. This way, I can get a handle on things and treat such coincidences as inevitable considering the large sample sizes that exist in the universe. By applying such rules, I can feel better about the world I live in. I can sleep peacefully in the arms of chance, and I have no need for mystical, magical explanations. Here are three strategies for tackling the next amazing coincidence you come across.

Compare the Number of Possible Outcomes

When I was a kid, I used to see a common advertisement in the comic books I read (e.g., *Statboy and His Flying Dog, Parameter*). The ad sold U.S. pennies that had been altered to include a portrait of John F. Kennedy in addition to the standard Lincoln profile. To justify why these two presidents should be included together, a long list of "remarkable" coincidences shared by these two presidents was presented (and, as I recall, if I purchased a set of these pennies, I would even get a small poster that listed these similarities.)

The list included things beyond the obvious, such as the facts that both were assassinated and both were succeeded by vice presidents named Johnson. I could (and did) interpret these coincidences as evidence of some important, somewhat-magical connection between the two. Let's use these coincidences as an example and approach it as a research question: is there an unusual number of similarities between these two presidents?

 It occurs to me now that the comic book ad led me to think for a time that the word *coincidence* was derived from the word *coin*. Of course, I quickly learned otherwise (by graduate school, certainly) that it has something to do with co-incidents.

One tool to use when deciding whether a coincidence is remarkable or predictable is to count the number of possible outcomes and then determine whether the given outcome (the coincidence) is unlikely to have occurred by chance. This is the approach taken when predicting shared birthdays in a large group [Hack #45].

Column one of Table 6-1 presents a list of some of the coincidences shown in those old comic book ads and also found in "Hard to Believe"-type publications. Column two shows a brief list of characteristics that both men could have shared, but did not.

Table 6-1. Comparing Abraham Lincoln and John F. Kennedy

Some amazing coincidences	Some unremarkable noncoincidences
Both assassinated.	Different heights.
Both elected in years ending with 60.	Different weights.
Kennedy assassin shot from a warehouse and hid in a theater. Lincoln assassin shot in a theatre and hid in a warehouse (well, a barn anyway).	They were different ages when they died (though they *were* the same age when they were born).
Lincoln was shot in Ford's theater. Kennedy was shot in a Ford.	They were born on different dates in different years.
Both were killed on a Friday.	Both men had different middle names.

Table 6-1. Comparing Abraham Lincoln and John F. Kennedy (continued)

Some amazing coincidences	Some unremarkable noncoincidences
Both were killed while sitting next to their wives.	Both men had wives with different names and, probably, different shoe sizes.
Both succeeded by men named Johnson.	Succeeded men with different names.
	Lincoln had a beard; Kennedy did not. (Come to think of it, their faces are different in hundreds of ways.)
	Kennedy probably had bowled occasionally; Lincoln never bowled a game in his life.

By paying attention to only the relatively few concordances between Lincoln and Kennedy (the hits) and ignoring all the non-hits, of which there are almost infinitely more, it is easy to misperceive the existence of some uncanny link. Of course, there still might be some uncanny link, but the "coincidences" do not provide evidence for it.

Figure Out the Actual Odds

If you play poker with any regularity (and, if you are a minor Hollywood celebrity, you apparently play all the time), you know that you rarely see a royal flush: a five-card hand with the 10, Jack, Queen, King, and Ace all of one suit. If your opponent were dealt a royal flush, would that be remarkable? Would you suspect cheating? It all depends on how many poker hands you have seen in your lifetime, I guess, or perhaps in recent memory.

Let's use a simple deal of five cards to do our math. To figure the chances of getting a royal flush on one deal of five cards, we would first calculate the number of possible five-card poker hands and compare that to the number of those combinations that are defined as a royal flush. The process takes three steps:

1. Calculate the number of possible hands, if order makes a difference. We start this way because the math is easiest. Any one card of 52 could be the first card, then any one of the remaining 51 could be next, then any one card out of 50, and so on down to any one card out of 48. So, the number of possible hands when the order matters is:

$$52 \times 51 \times 50 \times 49 \times 48 = 311,875,200$$

2. Order does not matter, though. So, we divide this giant total of all possible hands by the number of possible different sequences of cards. This number of different sequences is $5 \times 4 \times 3 \times 2 \times 1 = 120$, so the number of possible five-card poker hands is:

$$311,875,200 / 20 = 2,598,960$$

3. Because there are only four possible royal flushes, one for each suit, we divide this number of positive outcomes (4) by the total number of possible outcomes (2,598,960), for a probability of .000001539, or 1 out of 649,740.

Your opponent or you should be dealt five cards that make a royal flush once every 649,740 hands. So, if it does happen, it is certainly rare. If it happens more than once in the same game, you should interpret that as an amazing coincidence or as evidence of cheating. You decide. I know what my calculator and I would guess.

> What about drawing to a royal flush? After all, in draw poker and in Texas Hold 'Em, players have an opportunity to improve their hand or at least guide it toward some objective. In draw poker, if you have four cards to a royal flush and wish to discard the fifth and draw a new card, you have a 1 out of 47 chance for success, or .021 percent. If you have two chances to improve your hand, the odds go up to .043 percent, or about 1 out of every 25 attempts.

Remove Meaning Assigned to Meaningless Events

The human brain is at its best when it must make meaning out of data. Our remarkable intelligence can find meaning even where there is none. Often, this is the case when we think we have witnessed a miraculous set of coincidences. We see coincidences when we look for them.

Highly improbable events happen all the time—every day, and every minute of every hour. The highly improbable events are interesting only when we decide they are interesting. Think of our poker example. Because there are about 2.6 million possible five-card poker hands, the chances of any specific hand are one out of about 2.6 million. The odds are the same for the hands we have decided are particularly meaningful, such as a 10, Jack, Queen, King, and Ace of Spades, as they are for hands that we have decided are not particularly meaningful, such as a 4 of Clubs, 6 of Spades, Jack of Diamonds, Queen of Spades, and Ace of Hearts. Why is it amazing that you just drew a royal flush and not equally amazing when you draw any other random combination of cards? The probability is the same for all poker hands. We assign the meaning to a particular outcome.

The next time you are at a crowded place, such as a baseball game, amusement park, or airport, and you run into someone you know, notice that the coincidence is meaningful only because you happen to know the person. Yes, the chances were slim that you would run into that particular person (unless you are being stalked), but it is 100 percent certain that you would run into other people. All those other people just happen to be there the same time you are. It is a coincidence, and it is highly improbable that this particular mix of individuals is in the same place at the same time. It is not a meaningful coincidence for you, though.

> The odds are even good that you would run into someone you *know*, if we count *anybody* you know. Let's say you know 200 people and you, by yourself, go to a Kansas City Royals baseball game one night. If each of those 200 people goes to a Royals game one time each season and there are 81 home games, each of those 200 people has a 1/81 chance of being there the same night as you. It's unlikely then that you would run into any particular person—such as your Uncle Frank, for example—but it is highly likely that *someone* you know will be there. There is about a 92 percent chance that one or more of your 200 pals will be there, even though each of them rarely goes to a game. Even if you only know 56 people, the chances are greater than 50 percent that one or more of them will be there.

We are constantly exposed to a large set of events and people and things that interact and coincide in very unlikely ways. Occasionally, those coincidences have meaning to us, and so we notice them. What is amazing is that we do not notice these highly improbable events more often.

HACK #63 Sense the Real Randomness of Life

Before you accuse the casino of running a crooked game or threaten your boss with a lawsuit for hiring only blonde women, here's a tool for separating those nonrandom-seeming situations that probably did occur randomly from those nonrandom-seeming situations that probably did not occur randomly. Probably.

As you become more and more aware of the role that chance plays in the world around you, and begin to habitually stat-hack your way through everyday situations, you might become overly sensitive to patterns that don't seem right. Don't abuse your newfound powers, though, and treat probabilities as certainties. Additionally, don't make the mistake of expecting events that are supposed to be random to look random.

What Does Random Look Like?

Looking random and *being* random are not the same things. When events have several possible and equally likely outcomes, any of them can happen. The way the human mind works, though, many people think that the pattern of outcomes of events with several equally likely outcomes ought to look a certain way, a way that somehow looks random (whatever that means).

For example, real-world research has found that people tend to believe that, when flipping coins, the most probable outcomes are those that look the most mixed up. To illustrate this idea, look at Table 6-2. (Avoid looking at Table 6-3 until you have read a bit more.) Which exact sequence of coin flips do you think is most likely to occur?

Table 6-2. Coin-flip patterns, with probabilities not shown

Answer	Pattern of heads ands tails	Probability
A	Heads, Tails, Heads, Heads, Tails	?
B	Tails, Tails, Tails, Tails, Tails	?
C	Heads, Heads, Tails, Tails, Tails	?
D	Heads, Heads, Heads, Heads, Tails	?

Many people give the answer "A." Maybe you did, too. When asked to explain why A seems the most likely outcome, the answers include statements like these:

- "The others are too ordered."
- "A is more mixed up, so it's more likely."
- "A looks more random, like it could really happen."

Even though you know that coin flipping is random (assuming the coin isn't weighted), looking random doesn't make something more probable. All of these patterns of coin flips are actually equally probable, as shown by the math in Table 6-3.

Table 6-3. Coin-flip patterns, with probabilities

Answer	Pattern of heads and tails	Probability
A	Heads, Tails, Heads, Heads, Tails	$1/2 \times 1/2 \times 1/2 \times 1/2 \times 1/2 = 1/32 = .03125$
B	Tails, Tails, Tails, Tails, Tails	$1/2 \times 1/2 \times 1/2 \times 1/2 \times 1/2 = 1/32 = .03125$
C	Heads, Heads, Tails, Tails, Tails	$1/2 \times 1/2 \times 1/2 \times 1/2 \times 1/2 = 1/32 = .03125$
D	Heads, Heads, Heads, Heads, Tails	$1/2 \times 1/2 \times 1/2 \times 1/2 \times 1/2 = 1/32 = .03125$

When asked to predict a *specific* outcome of a series of coin flips, all possible outcomes must be equal, because each flip of the coin is independent of the other flips. In other words, the coin doesn't know whether it just landed on Heads or Tails, so there is no way that the coin can *know* which side it is supposed to land on the next time it is flipped. A coin, like dice or a roulette wheel, has no memory.

How to Spot Random Outcomes

To know an unusual sequence of events when you see it, you need to decide whether you are supposed to be paying attention to a *combination* or a *permutation*. In probability theory, we talk about calculating odds by looking at the probabilities of certain *combinations* (say, three Heads and two Tails in any order) and the probabilities of certain *permutations* (an exact sequence that would result in three Heads and two Tails, such as Heads, Tails, Heads, Heads, Tails, in that particular order).

If you are asked a question about which outcome is the most likely, or whether a given outcome could have occurred by chance, first determine whether you are being asked about combinations (the total number of Heads and Tails in any order, for example, or the number of different ways of drawing five playing cards of the same suit) or about the permutations that are possible. Here are the important distinctions between the two:

Combinations
> A combination is the total number of ways that one could end up with a particular number of values when drawing randomly from some population. Coin flips are samples drawn from a theoretically infinitely large population made up of 50 percent Heads and 50 percent Tails. The number of combinations varies, depending on the number of a certain value one is interested in. In other words, with five draws or flips, there are more ways to draw out three heads than there are ways to draw out five heads. So, drawing three heads is likelier than five heads.

Permutations
> Permutations are the number of ways that a given number of elements could be arranged. In other words, they are the number of exact sequences. In our coin-flip example, 5 elements that can each be 1 of 2 values results in 32 different possible orders of arrangement. So, each of the permutations shown in Table 6-3 will occur 1 out of every 32 times.

How to Calculate Combinations

The number of possible combinations is calculated by taking the number of possible values for one draw (e.g., two values for a coin: Heads or Tails) and multiplying it by itself for each draw:

$$\text{number of values}^{\text{number of draws}}$$

There are 32 possible combinations of 5 coin flips (2^5).

The equation for computing the *number of ways* to get a *particular* draw (e.g., three Heads) out of a particular number of elements drawn from a population is:

$$\frac{n!}{r!(n-r)!}$$

The previous equation requires these variables:

n The number of elements or draws (e.g., 5 coin flips).

r The particular draw of interest (e.g., 3 Heads).

!   Factorial, which means to take the number and multiply it by that number minus 1, then by that number minus 2, and so on, all the way down to 1. For example, 5! represents $5 \times 4 \times 3 \times 2 \times 1 = 120$ (which, by the way, is why there are 120 possible combinations of five cards in a poker hand [Hack #62]).

So, the number of ways to get three Heads out of five coin flips is:

$$\frac{5!}{3!(5-3)!} = \frac{120}{6(2!)} = \frac{120}{12} = 10$$

10 combinations out of 32 possible combinations means that you will get exactly 3 heads by flipping a coin 5 times 10/32 times, or about 31 percent of the time.

Statistics Hacking on a Desert Island

If you were on a desert island and didn't have access to books or equations and had to find out how often exactly three heads *should* come up in a group of five coin flips, you could use the brute force method of listing all the possible patterns of flips and counting how many of them have exactly three heads. It would look like this, with the outcome of interest (three heads) shown in bold:

HHHHH THHHH HHHHT **THHHT HHTTH** THTTH HHTTT THTTT
HHHTH **THHTH HHHTT** THHTT HHTHH **THTHH HHTHT** THTHT
HTHHH **TTHHH HTHHT** TTHHT HTTTH TTTTH HTHTT TTHTT
HTTHH TTTHH HTTTT TTTTT **HTHTH** TTHTH HTTHT TTTHT

When to Be Suspicious

Deciding whether a pattern is random (i.e., what one would expect by chance) is a matter of:

- Knowing the chances of certain combinations (not permutations)
- Fighting the psychological tendency to expect chance results to not produce a recognizable pattern
- Setting a standard for how unlikely an event must be before questioning the data

Let's return to our table of coin flips, shown now in Table 6-4 with the added chances of certain outcomes of interest.

Table 6-4. Coin-flip outcomes and probabilities

Order	Order probability	Outcome	Outcome probability
Heads, Tails, Heads, Heads, Tails	.03125	Three Heads	.31250
Tails, Tails, Tails, Tails, Tails	.03125	Five Tails	.03125
Heads, Heads, Tails, Tails, Tails	.03125	Three Tails	.31250
Heads, Heads, Heads, Heads, Tails	.03125	Four Heads	.15625

The rarest of these outcomes is five Tails, which will occur about 3 times for every 100 times you produce five coin flips. It is unlikely to happen by chance on a given attempt, but it will happen occasionally across a series of attempts. If it happens frequently across a series of attempts, something might be up.

What level of likelihood are you comfortable with? How rare must an event be before you decide it did not occur by chance? Scientists have set a standard of 5 percent. If study results suggest an outcome that would occur by chance only 5 percent or less of the time, it is usually considered to be significant, and is probably evidence that something other than chance is in play.

You get to decide for yourself, though, when you want to accuse someone of being a cheat. Good luck on making that decision! It should result in fist fights less than 5 percent of the time.

—*Jill Lohmeier with Bruce Frey*

HACK #64   Spot Faked Data

If you haven't given it much thought before, it might be quite natural to assume that all digits are equally likely to show up in most random data sets. But according to Benford's law, for many types of naturally occurring data, the lower the digit, the more frequently it will occur as a leading digit. You can use this secret knowledge to check the authenticity of any data set.

In the 19th century, long before the age of electronic calculators, scientists used tables published in books to find values of logarithms. A particularly observant 19th-century astronomer and mathematician, Simon Newcomb, noticed that the pages of logarithm tables were more worn in the first pages than in the last pages. Newcomb concluded that numbers beginning with 1 occur more frequently than numbers beginning with 2, numbers beginning with 2 occur more frequently than numbers beginning with 3, and so on.

Newcomb published an empirical result based on his observations in the American Journal of Mathematics in 1881, which stated the probabilities of a number in many types of naturally occurring data, beginning with digit d for $d = 1, 2, \dots 9$. Newcomb's *first significant digit law* received little attention and was largely forgotten until over 50 years later when Frank Benford, a physicist at General Electric, noticed the same pattern of wear and tear of logarithm tables.

After extensive testing (20,229 observations!) on a wide variety of data—including atomic weights, drainage areas of rivers, census figures, baseball statistics, and financial data, among other things—Benford published the same probability law concerning the first significant digit in the Proceedings of the American Philosophical Society (Benford, 1938). This time, the first significant digit law attracted greater attention and became known as *Benford's law*. Although Benford's law became fairly well known after the 1938 paper, which included substantial statistical evidence, it lacked a rigorous mathematical foundation until that evidence was provided by Georgia Tech Mathematics professor Theodore Hill in 1996 (Hill, 1996).

Today, Benford's law is routinely applied in several areas in which naturally occurring data arise. Perhaps the most practical application of Benford's law is in detecting fraudulent data (or unintentional errors) in accounting, an application pioneered by Saint Michael's College Business Administration and Accounting professor Mark Nigrini (*http://www.nigrini.com/*).

The detection of fabricated data is important not only in accounting, but also in a wide variety of other applications (for example, clinical trials in drug testing). This hack describes Benford's law, shows you how to apply it, provides some intuitive justification on why it works, and gives some guidelines on when Benford's law can be applied.

How It Works

In its simplest form, Benford's law states that in many naturally occurring numerical data, the distribution of the first (nonzero) significant digit follows a logarithmic probability distribution described as follows. Following Hill (1997), let $D_1(x)$ denote the first base 10 significant digit of a number x. For example, $D_1(9108) = 9$, and $D_1(0.025108) = 2$.

Then, according to Benford's law, the probability that $D_1(x) = d$, where d can equal 1, 2, 3, ..., 9, is given by the following equation:

$$P(D_1{=}d) \;=\; \log_{10}\!\left(1 + \frac{1}{d}\right)$$

Thus, Table 6-5 gives the probabilities of the first significant digits.

Table 6-5. Probabilities of first digits under Benford's Law

First nonzero digit	Probability according to Benford's law
1	0.301
2	0.176
3	0.125
4	0.097
5	0.079
6	0.067
7	0.058
8	0.051
9	0.046

Laying Down the Law

To demonstrate Benford's law, I'll consider two examples that you can verify yourself.

Street addresses. To see Benford's law in action, open the phone book of your city or town to any page, and record the number of house numbers that begin with each nonzero decimal digit. Two pages should be sufficient. Unless there is something very unusual about your town, the relative frequencies should resemble the respective probabilities predicted by Benford's law.

Table 6-6 shows results computed from the 413 house numbers taken from two pages of the 2005–2006 Narragansett/Newport/Westerly, RI Yellow Book (White Pages section).

Table 6-6. Addresses following Benford's law

First nonzero digit	Relative frequency for first digit of house number	Probability according to Benford's law
1	0.334	0.301
2	0.174	0.176
3	0.143	0.125
4	0.075	0.097
5	0.073	0.079
6	0.075	0.067
7	0.046	0.058
8	0.043	0.051
9	0.036	0.046

Figure 6-1 shows the pattern more clearly.

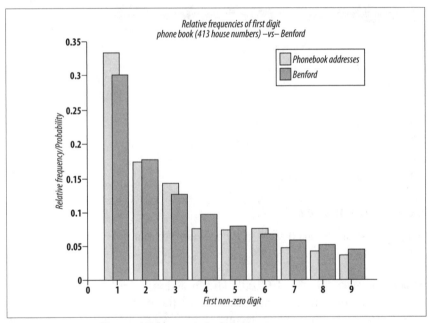

Figure 6-1. Street addresses following Benford's law

Although the agreement with Benford's law is not perfect, you can see a reasonably good fit. If you take a larger sample of addresses, the resulting relative frequencies will be even closer to the probabilities predicted by Benford's law.

Stock prices. The stock market is known to follow Benford's law. You can verify this yourself by obtaining up-to-the-minute NASDAQ Securities prices at *http://quotes.nasdaq.com/reference/comlookup.stm*.

Figure 6-2 and Table 6-7 show the relative frequencies of the first nonzero decimal digits for NASDAQ Securities as of January 27, 2006, compared to the probabilities predicted by Benford's law.

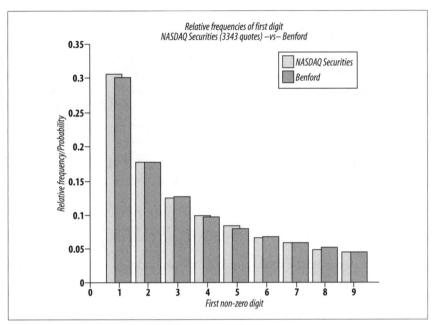

Figure 6-2. The stock market following Benford's law

Table 6-7. NASDAQ securities following Benford's law

First nonzero digit	Relative frequency for first digit of NASDAQ securities	Probability according to Benford's law
1	0.301	0.301
2	0.167	0.176
3	0.133	0.125
4	0.095	0.097
5	0.082	0.079
6	0.071	0.067
7	0.055	0.058
8	0.045	0.051
9	0.049	0.046

You can obtain the Matlab code used to produce the tables and figures in this section at *http://homepage.mac.com/samchops/benford/*. Additionally, Mark Nigrini provides his DATAS software (including a free student EXCEL program), which performs a more sophisticated data analysis of the first, second, and first two digits, at *http://www.nigrini.com/datas_software.htm*.

More General Statements of Benford's Law

Benford's law does not apply to the first nonzero digit only, but also includes probabilities of other digits. Once again, following the treatment discussed earlier, let $D_2(x)$ denote the second base-10 significant digit of a number x. For example, $D_2(9108) = 1$, $D_2(9018) = 0$, and $D_1(0.025108) = 5$. Notice that, unlike the first significant digit, the second significant digit can be zero.

Then, according to Benford's law, the probability that $D_2(x) = d$, where d can equal 0,1, 2, ..., 9, is given by the following equation:

$$P(D_2=d) = \log_{10}\left[1 + \left(\sum_{i=1}^{k} d_i \cdot 10^{k-1}\right)^{-1}\right]$$

This formula leads to the probabilities of the second significant digit, shown in Table 6-8.

Table 6-8. Benford's second-digit law

Second significant digit	Probability according to Benford's law
0	0.11968
1	0.11389
2	0.10882
3	0.10433
4	0.10031
5	0.09668
6	0.09337
7	0.09035
8	0.08757
9	0.08500

From Table 6-8, you can see that the differences among the probabilities of the second digit are not nearly as dramatic as those probabilities corresponding to the first digit.

Now, back to the stock market. To illustrate Benford's law as it relates to the second significant digit, I computed the relative frequencies of the second significant digits of our earlier NASDAQ Securities example. The results in Table 6-9 show, again, a close agreement with Benford's law.

Table 6-9. NASDAQ securities following Benford's second-digit law

Second digit	Relative frequency of second digit	Probability according to Benford's law
0	0.12803	0.11968
1	0.11427	0.11389
2	0.10918	0.10882
3	0.10290	0.10433
4	0.10230	0.10031
5	0.09273	0.09668
6	0.09064	0.09337
7	0.09153	0.09035
8	0.08406	0.09035
9	0.08436	0.08500

A more general Benford's probability formula can be used to compute the respective probabilities of the nth digit. Let $D_k(x)$ denote the kth base-10 significant digit of a number x. Then, according to Benford's law, the probability that $D_1(x)=d_1$, $D_2(x)=d_2$,..., and $D_n(x)=d_n$ is given by the following equation:

$$P(D_1=d_1, D_2=d_2, \ldots, D_n=d_n) = \log_{10}\left[1 + \left(\sum_{i=1}^{n} d_i \cdot 10^{n-1} \right)^{-1} \right]$$

Note that if k does not equal 1, then d_k can equal 0, 1, 2, ..., 9 and, as noted earlier, d_1 can equal 1, 2, ..., 9.

Where Else It Works

Two unique properties of Benford's Law are *scale invariance* and *base invariance*.

Scale invariance. Benford's law is scale-invariant; that is, if you multiply the data by any nonzero constant, you still wind up with a distribution that closely follows Benford's law. Thus, it makes no difference whether you measure stock quotes in dollars, dinars, or shekels, or whether you measure lengths of rivers in miles or kilometers. You'll always wind up with data that follows Benford's law.

Spot Faked Data

To prove this, I took the NASDAQ securities data used in the earlier example and multiplied each value by π. As you can see in Table 6-10, the relative frequencies still follow Benford's law.

Table 6-10. NASDAQ securities scaled by π following Benford's law

First nonzero digit	Relative frequency for first digit of NASDAQ securities	Probability according to Benford's law
1	0.306	0.301
2	0.176	0.176
3	0.123	0.125
4	0.097	0.097
5	0.081	0.079
6	0.066	0.067
7	0.058	0.058
8	0.049	0.051
9	0.045	0.046

Base invariance. The base-invariant property of Benford's law states that it applies not only in base 10, but also in more general bases. Moreover, Theodore Hill showed that Benford's law is the only probability law that has this property (Hill, 1995).

> You can find the formula for Benford's law in the general base-b case in Hill (1997). See the "See Also" section for publication details.

Benford's law works best on data that has the following characteristics:

Sufficient variability
    The higher the variability, the better Benford's law applies.

No built-in maximum or other similar constraint
    For example, Benford's law does not apply to the ages of high school seniors, or to members of the local senior citizen center.

Numbers that result from counting or measuring
    For example, it does not work well for social security numbers and ZIP Codes, because they are simply identifiers and are not true numerical values.

Large sample size
    The larger the data set, the better Benford's law applies.

Random sampling
> The data results from a large number of random samples from a large number of randomly selected probability distributions. This realization by Hill led him to his proof of Benford's law (Becker, 2000; Hill, 1999).

Since tax data strongly follows Benford's law, it has been used quite successfully to identify fraudulent tax returns. In describing some of the basic features of Benford's law, we showed how anyone can perform a quick-and-dirty test for irregularities in data. Specifically, anyone can easily compute relative frequencies of first digits and eyeball the results juxtaposed with probabilities predicted by Benford's law.

In practice, the programs used by experts and authorities to identify deviations from Benford's law and other irregularities can be quite sophisticated. It is also important to keep in mind that deviation from Benford's law does not prove fraud, but it does raise red flags suggesting that further investigation might be indicated.

> For more details on the application of Benford's law to detect fraud, including a "goodness-of-fit" test, see Nigrini (1996). Consult this hack's "See Also" section for publication details.

Why It Works

Although the proof of Benford's law is quite technical, there are some insightful and intuitive explanations for this mathematical principle. One such explanation that I find particularly attractive has been provided by Mark Nigrini (1999).

His explanation goes something like this. If you imagine that some investment with an initial amount of $100 is expected to grow at an annual rate of 10 percent, it would take about 7.3 years for the first digit of the total amount to change to 2. This is because the total amount has to increase by 100 percent to reach a value of $200. In contrast, consider the time it would take for $500 to increase to $600. If we continue to assume an annual growth rate of 10 percent, it would take about 1.9 years to reach $600. So, the amount of time until the investment amount has a first digit of 5 is considerably less that the amount of time it has a first digit of 1. Once the total amount reaches $1,000, it will again take about 7.3 years before it will have a first digit of 2 (after another 100 percent increase).

The real world is a bit more complicated, but this does help to explain why 1 is a more common first digit than larger digits. Another intuitive explanation is that there are more small towns than large cities, and there are more short rivers than long rivers.

Where It Doesn't Work

Benford's law is less likely to apply in data sets with insufficient variability or data sets that are nonrandomly selected. For example, computer files sizes approximately follow Benford's law, but only if no restriction is placed on the type of files selected.

To illustrate this, I found the frequencies of the first digit of the file sizes on an Apple PowerBook G4. The results shown in Figure 6-3 and Table 6-11 exhibit the Benford's law pattern.

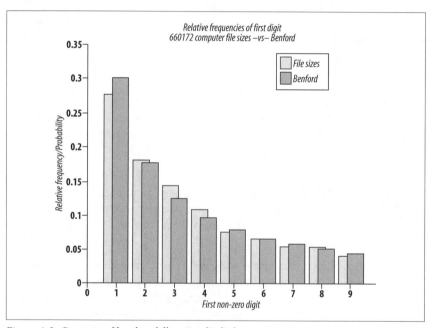

Figure 6-3. Computer files that follow Benford's law

Table 6-11. Computer files that approximately follow Benford's law

First nonzero digit	Relative frequency for first digit of 660,172 computer files	Probability according to Benford's law
1	0.277	0.301
2	0.181	0.176
3	0.144	0.125
4	0.107	0.097
5	0.076	0.079
6	0.067	0.067

Table 6-11. Computer files that approximately follow Benford's law (continued)

First nonzero digit	Relative frequency for first digit of 660,172 computer files	Probability according to Benford's law
7	0.054	0.058
8	0.054	0.051
9	0.041	0.046

Although the results shown in Figure 6-3 and Table 6-11 are based on 660,172 files, Table 6-12 demonstrates that a sample size of 600 is large enough to exhibit the Benford's law pattern (albeit not as well as the larger sample), provided the sample of files is random.

Table 6-12. Random selection of 600 computer files sizes

First nonzero digit	Relative frequency for first digit of 600 computer files	Probability according to Benford's law
1	0.262	0.301
2	0.187	0.176
3	0.147	0.125
4	0.107	0.097
5	0.069	0.079
6	0.070	0.067
7	0.052	0.058
8	0.057	0.051
9	0.052	0.046

For comparison, I computed the relative frequencies of MP3 files in an iTunes music library on the same computer. Table 6-13 and Figure 6-4 show that this set of files does not follow Benford's law.

Table 6-13. Music MP3 files that do not follow Benford's law

First nonzero digit	Relative frequency for first digit of 601 MP3 files	Probability according to Benford's law
1	0.080	0.301
2	0.097	0.176
3	0.276	0.125
4	0.270	0.097
5	0.161	0.079
6	0.070	0.067
7	0.023	0.058
8	0.013	0.051
9	0.001	0.046

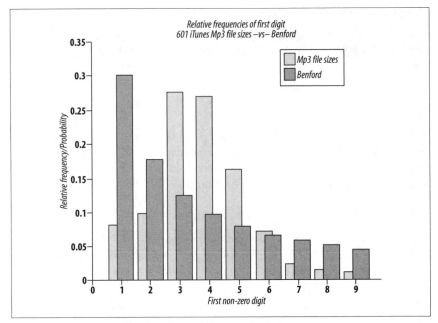

Figure 6-4. Music MP3 files that do not follow Benford's law

The fact that the file sizes of about 600 MP3 music files do not approximate Benford's law is not surprising, since the sizes of MP3 music files exhibit much less variability than a more random selection of any 600 computer files.

See Also

- Becker, T. J. (2000). "Sorry, wrong number: Century-old math rule ferrets out modern-day digital deception," *Georgia Tech Research Horizons, http://gtresearchnews.gatech.edu/reshor/rh-f00/math.html.*

- Browne, M. (1998). "Following Benford's law, or looking out for no. 1." *The New York Times*, August 4, 1998.

- Fawcett, W. (n.d.). "Significant figure generator." *http://williamfawcett. com/flash/SigFigDistbGen.htm.*

- Benford, F. (1938). "The law of anomalous numbers." *Proceedings of the American Philosophical Society, 78, 551–572.*

- Hill, T. P. (1996). "A statistical derivation of the significant digit law." *Statistical Science, 10, 354–363.*

- Hill, T. P. (1995). "Base-invariance implies Benford's law." *Proceedings of the American Mathematical Society, 123, 887–895.*

- Hill, T. P. (1997). "Benford's law." *Encyclopedia of Mathematics Supplement, 1, 112.* Kluwer.

- Hill, T. P. (1999). "The difficulty of faking data." *Chance,* 26, 8–13.

- Newcomb, S. (1881). "Note on the frequency of use of the different digits in natural numbers." *American Journal of Mathematics,* 4, 72–40.

- Nigrini, M. (1999). "I've got your number: How a mathematical phenomenon can help CPAs uncover fraud and other irregularities." *AICPA Journal of Accountancy Online Journal,* May 1999, *http://www.aicpa.org/pubs/jofa/may1999/nigrini.htm.*

- Nigrini, M. (1996). "A taxpayer compliance application of Benford's law." *Journal of the American Taxation Association,* 18, 72–91.

- You can obtain the *Matlab* code used to produce the tables and figures in this section at *http://homepage.mac.com/samchops/benford/.* You'll need to have *Matlab* (*http://www.mathworks.com*) installed to run the code.

—*Ernest E. Rothman*

Give Credit Where Credit Is Due
#65

Stylometrics is a statistical procedure that identifies the underlying dimensions that define an author's style. It uses the method of factor analysis to judge who wrote what.

Professor Howe-Mutch had a problem. Two of his best students were sitting in his office, hoping to resolve a dispute. Dr. Howe-Mutch had awarded an A+ to Paul's final paper (on the historical importance of chocolate milk). The problem was that Lisa claimed to have written it. An accusation of plagiarism had been made! Both were good students who had written many quality papers for him in the past. So, the solution as to true authorship was not a simple one, nor was the realization that one of his favorite students was a cheat.

Fortunately, the good doctor of philosophy had many years of experience and was wiser than his adjunct position at State Community College and Trucking School might have suggested. Among other obscure statistical hobbies, Dr. Howe-Mutch dabbled in the art of *stylometry*, a statistical method for categorizing the style of written works. The method can also be used to identify anonymous authors. It works best when there are a couple of possibilities or suspects to choose from, and when the typical writing styles of the suspects are known and have been quantified. Let's watch as the brokenhearted professor applies these techniques to find the true author.

Building a Model

First, Dr. Howe-Mutch asks Paul and Lisa to bring in all the other papers they have each written in the past and about which there is no dispute. In just a few moments, the papers are scanned into a computer, providing a database of all the different words used by both writers.

Or they were sent to him electronically so no scanning was necessary; none of this is relevant to the story, so why are you questioning me about it?

For the first analysis, all the words written by the two writers are kept together. Dr. Howe-Mutch counts the frequency with which each word is used and identifies the 50 to 100 most commonly used words in the combined database. These words become the items or key variables that supply the data for a *factor analysis*. Factor analysis is a statistical process that looks at the correlations [Hack #11] among groups of variables and identifies clusters of variables that correlate better among themselves than they do with other variables. Whatever these grouped-together variables have in common is assumed to be a factor, component, or dimension that they all share.

For the sake of our story, I'll show only 10 of the words that Dr. Howe-Mutch identified as most common across both writers' works. Table 6-14 shows the words and their frequency of use. When looking at all the words Paul and Lisa wrote, *the* was used 4.2 percent of the time, *weasel* was used 1 percent of the time, and so on.

Table 6-14. Paul and Lisa's commonly used words and their frequency

Word	Frequency
the	4.2 percent
and	2.1 percent
to	1.8 percent
a or an	1.2 percent
weasel	1.0 percent
of	0.8 percent
in	0.8 percent
that	0.5 percent
it	0.4 percent
not	0.2 percent

These words act as variables to try to identify the underlying factors that describe one or more dimensions of style. Paul and Lisa's styles might be at different places along these dimensions. It might be that only one dimension or factor is necessary to account for variability in the usage of these words, or there might be many dimensions. Once these dimensions— defined by the variables that correlate together, or *load,* on the dimension— are identified, any writing sample could be *placed* in the theoretical space framed by the factors.

The data for Dr. Howe-Mutch's factor analysis are supplied by each section of 500 words in the writing samples. Each section receives a score on each of the word variables. The score will be the number of times the word is used in that paragraph. Table 6-15 shows examples of the data Mr. Howe-Mutch collects.

Table 6-15. Sample of study data

	the	and	to	a/an	weasel	of	in	that	it	not
Section 1	21	8	11	5	4	0	0	1	0	2
Section 2	10	7	15	5	2	10	1	0	0	0
Section 3	5	5	5	2	6	12	2	4	1	0
Section 4	0	2	4	3	1	4	6	8	1	0
Section 5	4	11	16	2	0	3	5	0	3	1

 In Table 6-15, scores indicate the number of times each word appears in the text sections.

Factor Analysis

Next, Dr. Howe-Mutch performs the factor analysis, a fairly complex mathematical process that these days is done using computers, while the researcher makes many theory-driven decisions at different points along the way. Basically, the factors are identified by exploring the relationships among variables until a small number of variable groupings are found that seem to account for as much variability as possible across the data. The commonality shared by variables in each grouping provides the mathematical fodder that defines the factor. Once the factors are chosen, any observation—in this case, a sample of text—can receive scores on the factor and then be placed in that theoretical space, with the factor scores serving as coordinates.

In this case, the analysis suggests that two factors do a good job of describing the sample texts. Factor 1 is defined by the use of words such as *the* and *a/an* at one end and *of* and *in* at the other. In other words, the text sections differed based on how frequently they used articles, and the sections that had a higher frequency of article use tended to be lower in their use of prepositions. Factor 2 is defined by the frequency of the use of the word *weasel*.

In exploratory factor analysis, typically researchers are interested in discovering and naming the underlying *constructs* (i.e., invisible traits) that account for human behaviors and characteristics. For this use, though, Professor Howe-Mutch is interested only in defining these dimensions based on the variables (e.g., word use) that anchor them at both ends. He is not interested in figuring out why those text sections that tend to contain the word *the* frequently also tend to contain *a* or *an* frequently. He also is not interested in why the use of the word *weasel* could distinguish between his different writing samples. For his purposes, he is content just knowing that these two factors provide a couple of good axes to map the space of all words that the two authors chose to use in their samples.

When the factor scores from each of Paul and Lisa's sample papers are computed, it becomes clear that the two authors have different styles. Lisa tends to use the word *weasel* more frequently than Paul; her papers score high on Factor 2. Lisa's papers also tend toward the high use of articles and receive fairly high Factor 1 scores. Paul's papers, on the other hands, tend to avoid the use of the word *weasel* and tend toward the preposition end of Factor 1.

This is difficult to grasp using words alone, so an illustration will help draw a picture to demonstrate the placement of the sample texts. Figure 6-5 shows the two factors, the word usage that defines them, and where the different writing samples loaded on the two factors. For the sake of convenience for this discussion, Figure 6-5 displays only a few of the writing samples and maps only the 10 words in Table 6-14 and Table 6-15. Also included in the figure is the placement of the disputed paper in this theoretical dimensional space.

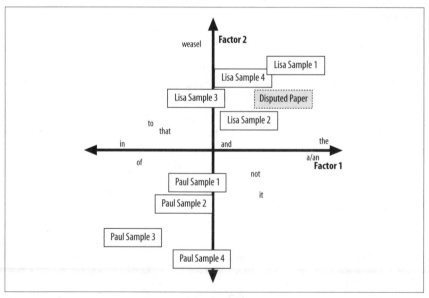

Figure 6-5. Factor analysis of text samples

The solution to the mystery is now clear. The disputed paper shares the characteristics of Lisa's papers, not Paul's. Because Paul and Lisa's earlier papers display a consistent but different style, at least as defined by word counts, the factor map is a useful tool to identify the most likely author of the paper.

Dr. Howe-Mutch awards the A+ to Lisa, accuses Paul of plagiarism, and is now engaged in a lengthy court battle with Paul's attorneys, which will no doubt leave our fine statistician friend penniless. The important thing, though, is that a statistical procedure was able to make the invisible visible. Science triumphed once again.

See Also

- "Who wrote the 15th book of Oz?," by J.N.G. Binongo in *Chance*, 16, 2, 9–17.

Play a Tune on Pascal's Triangle
Need to know the odds quickly? Pascal's Triangle is a simple layout of numbers that allows for quick and easy calculations of probability. It's worked for 300 years, so I bet it will work for you.

The thing that statisticians do most often is calculate probabilities, which can describe expected outcomes for a variety of situations. A simple example is flipping a coin. Imagine that you have been asked to wager on the outcome of a coin flip. With two possible outcomes, heads or tails, the chances of getting either outcome on a single coin flip is 1 out of 2, or 1/2.

The math is easy if you know the number of different ways to get the winning outcome and the number of possible outcomes. In the coin flip example, there's only one way to get a winning outcome, and there are only two possible outcomes. The math is just a bit harder if we have more than one coin flip and wonder about the number of all possible outcomes and how many of those combinations would match our winning criteria. For example, if I want two heads in a row on two coin flips, I could list all possible outcomes, identifying the number of those outcomes that make me a winner, and then see what proportion of all outcomes are winners for me. That proportion would be my chances of winning.

The number of possible outcomes that count as winners is often more complex than our simple coin flip examples, though, because there might be many trials (or dice rolls, or purchase of lottery tickets, or whatever) and many different combinations. For example, you might want to figure the number of possible combinations of different elements from any set of objects you're drawing out of a hat or choosing through some other random selection process.

Imagine you are one of six relatives driving to the airport and you must all drive there in one big van. None of you like each other much, so you need some fair way to decide who will sit where. You will randomly pick two names to drive together in the front seat.

> A private note to my Uncle Frank: yes, this example *is* based on the "unpleasantness" last Thanksgiving. All is forgiven, at least on my side of the family, but we agree it would be best if you brought your own car next year.

Now, you need to know the chances that you will be in the front seat and who you might be with. The problem is calculating how many different combinations of relatives could be in that front seat. For both simple wagers, such as coin flips, and life-and-death situations, such as long car trips, you can use a layout of numbers called Pascal's Triangle to do the math for you.

Presenting Pascal's Triangle

Pascal's Triangle is shown in Figure 6-6. This layout of numbers has some interesting properties. Here it is shown made up of 10 rows, with 10 numbers in its lowest row, but it can be made infinitely large with an infinite number of rows. The outer edges going down are all 1s. The next diagonals start with 1 but increase by 1 as they go down.

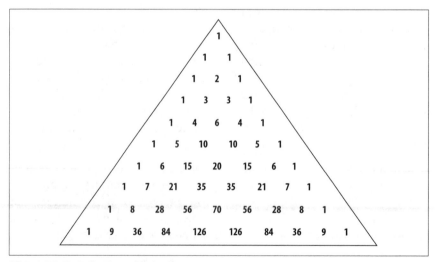

Figure 6-6. Pascal's Triangle

Similar interesting progressions are found throughout the Triangle. Notice that each number is the sum of the two numbers above it: 84 is 56 + 28, 7 is 6 + 1, and so on. These cool patterns aren't the reason that the Triangle is of interest to us, however. Instead, we're going to use it to calculate the probabilities for a variety of outcomes.

Calculating Probability Using the Triangle

Pascal's Triangle, named for Blaise Pascal (a very smart, very early contributor to probability theory who lived in the 1600s), has already made use of all the calculations we need to answer a variety of questions.

Though this pattern of numbers is known as *Pascal's Triangle*, Pascal did not, and never claimed to have, originated it. Similar patterns of numbers were presented by Pascal's teacher, Hérigone, as well as other colleagues writing at about the same time.

A general formula exists to determine the number of possible outcomes of a certain type. This general formula works any time there are exactly two possible outcomes; thus, the term *binomial coefficient* is used to describe the outcome of the formula (*bi-nomial* means *having two names* or, in a statistical sense, *two outcomes*). To determine the number of possible binomial combinations of outcomes across a given number of trials, this formula is used:

$$\text{Number of Possible Winning Combinations} = \frac{k!}{k!(n-k)!}$$

A range of possible values that could be entered into this formula are the coordinates on Pascal's map. The n in the equation, which represents the number of trials or events, indicates which row to go to. The k in the equation tells us which entry on that row to go to. The 1s along the left of the map are like a border: they count as zero. So, to use the triangle, we start counting at 0.

The exclamation point after some numbers in this formula means *factorial*, which, in turn, means that you are supposed to count down from that number to 1, multiplying the numbers together as you go. For example, *5 factorial* means $5 \times 4 \times 3 \times 2 \times 1$, or 120. By the way, by rule, 0! counts as 1.

Assessing the probability of flipped coin outcomes. For our second, slightly more complex coin flip question, the chances of getting exactly two heads when flipping a coin two times, use the triangle like this:

1. The row to go to is determined by the number of coin flips we will make: 2. The entry we will count over to in that row is determined by the number of heads we want to see: 2. With our coin flip example, 2 heads on 2 trials, count down two rows to this row:

   1 2 1

2. Then, count over two entries to the **1**. Our answer is 1, and so there is one chance that we will get two heads.

3. But 1 chance out of how many chances? Add up all the numbers in the row you are in to get that answer! 1 + 2 + 1 = 4, so our chances are 1 out of 4, or 25 percent.

The Triangle answers more complicated question, as well. Suppose you want *exactly* 3 heads out of six coin flips:

1. Count down six rows (remember to count the top of the triangle as 0). You get to this row:

   1 6 15 20 15 6 1

2. Count over three numbers and you hit 20. There are 20 different ways that exactly three heads could come up in six coin flips.

3. 20 out of how many possibilities, you ask? Summing all the values in that row gives us 64. 20 out of 64 times you will get exactly three heads (or three tails). That's about 31 percent of the time.

Assessing the probability of a bad car trip. Another way to use the triangle is to see how many combinations are possible from a certain number of elements drawn out in a certain way. Our car trip example is interested in how many possible combinations of two people can be drawn from a group of six.

From a set of six elements, you will draw out two of them and match those up. For this question, and in the binomial formula that defines the triangle, think of the six relatives as n and the two names to be drawn as the k:

1. Count down six rows and across two entries, and you hit the number 15. There are 15 possible combinations of two people drawn from six people.

2. In this case, you are interested only in your chances of being in the front seat with one specific person. That's 1 combination of front seat passengers out of 15 possible combinations. So, you will be matched up with your annoying Uncle Frank, or Aunt Tillie, or whomever, in the front seat just 1 out of 15 times.

Why It Works

The numbers in the triangle match the values that you would derive if you actually did the math using the binomial equation, but you'll notice that the triangle answers several other questions along the way. The patterns of the numbers, their progression, are consistent with other formulas used in determining probability.

For instance, the total possible number of coin flip combinations for six coin flips is answered on the triangle by totaling the values in the sixth row: 64. You would have mathematically derived that value by applying the general formula for number of possible outcomes for a coin: $2^{\text{number of flips}} = 2^6 = 64$.

As for the chances that you would both be chosen as one of two people out of six and that the other person would be a specific one of the other people (our trip to the airport example), the triangle said 1 out of 15. But you also could have figured it this way:

1. Chance of being in a group of two people out of six = 2/6 = .33
2. Chance of a specific "other" person being chosen = 1 person out of 5 "others" = 1/5 = .20
3. Chance of both outcomes occurring = .33×.20 = .066, and .066 = 1 out of 15

So, when you have a complicated-looking problem that involves combinations and permutations and so many possibilities that it makes your head spin, let the soothing music of Pascal's Triangle bring peace to your troubled mind.

Control Random Thoughts

The rambling nature of our inner thoughts is often perceived as creating an unpredictable random path. You can take advantage of this misperception to guess the thoughts of those around you by increasing the probability that they will focus on whatever you wish.

No stranger to creepy scenes, Edgar Allen Poe relates this one in *Murders in the Rue Morgue*:

> Occupied with thought, neither of us had spoken a syllable for fifteen minutes at least. All at once, Dupin broke forth with these words: "He *is* a very little fellow, that's true, and would do better for the *Theatre des Varietes*." "There can be no doubt of that," I replied unwittingly... "Dupin," said I gravely, "this is beyond my comprehension. I do not hesitate to say I am amazed, and can scarcely credit my senses. How was it possible you should know (what) I was thinking of...?"

Have you ever been talking to someone, and your mind wanders off for a little while? Then, you bring up whatever it was that you were thinking about and, lo and behold, the other person was thinking about the exact same thing!

Why does this happen? Can you make it happen? Can you predict what the other person is going to say? More than likely, yes, you can sometimes make it happen, and sometimes you can predict what the other person is going to say. This is especially true if the two of you share a common background.

Mind Control

Our memories are filled with words, thoughts, stories, and so on that are associated with other words, thoughts, and stories. If you want someone to think of a certain topic so that you can read her mind, the easiest way to trick her into thinking what you want her to think is by bringing up a topic that is closely related to the desired topic.

For example, if you want your friend to start thinking about lions and tigers and bears, you might *prime* her thought process with words that are associated with that theme—words such as *Wizard of Oz*, *Dorothy*, *Toto*, or even *stripes*, since stripes and tigers are highly associated with each other.

All words have a certain frequency of occurrence in written and spoken language. Some words have a very high frequency of occurrence (such as *the*, *it*, etc.), while other words have very low frequency of occurrence (such as *aardvark*). Additionally, some words occur with other words quite frequently (such as *salt and pepper* or *rhythm and blues*). In fact, they occur so often with the other words that research has found that people think both words even when only one is said.

By learning these associations, we can process incoming information more quickly. If we hear *salt* and are already thinking *salt and pepper*, we are one step ahead and can begin to reach for both before our dinner companion even finishes asking us to pass them.

So, if you want to "control" someone's mind, the trick is simply to know which things occur most frequently together. The more frequent a word is, the more likely someone is to think it. Likewise, the more frequently two words occur together, the more likely one is to think of both words when only one is stated.

Probability and Word Association

Researchers interested in those words that tend to be associated have collected data over the years to see what is normal for us humans. Psychiatrists use knowledge of typical free associations between words as a tool for reading the subconscious. Cognitive psychologists use the same information to map the way the brain processes information.

A huge amount of information is known about *cues* (the word presented that might lead to an association) and *targets* (the words thought of after the cue is presented). Table 6-16 shows a sample of word cues and the probability that normal people, such as your friends, will think of particular targets. The table provides a range of good cues and bad cues to give you an idea of how most minds work.

Table 6-16. Chances of word associations

Cue	Target	Probability
condom	sex	.53
bumpy	sex	.01
broccoli	green	.25
broccoli	gross	.01
pajamas	sleep	.36
accident	car	.36
accident	oops	.01
mother	father	.60
mother	goose	.02
orthodontist	teeth	.42
hero	Superman	.17
hero	Batman	.02
statistics	numbers	.26
statistics	boring	.03
coleslaw	fish	.01

Information like this is useful for when you want your subject to think of certain words or ideas. With *sex*, for example, you will have more luck cueing with *condom* than you will with *bumpy*.

> Table 6-16 draws on the seemingly exhaustive list of typical associations for thousands of words found at *http://w3.usf.edu/ FreeAssociation/*, provided by Nelson, McEvoy, and Schreiber, researchers at the Universities of South Florida and Kansas.

Building a List of Word Associations

Associated ideas and words form slightly different webs of connections in each person, but within groups of people with a shared culture (pop or otherwise) and shared experiences, the networks are similar. To be able to start saying your friends' thoughts out loud (and spooking the heck out of them), you'll need to know the likely associations in your metaphorical corner of the world.

You can conduct a small study to help you determine which words among your friends are most strongly associated with each other. Create a sample of a few representative friends or family members. Make up a list of test words, and ask your sample to say the first thing that comes to mind when you say each word. Words in common phrases or titles work well. Words that elicit thoughts of favorite in-jokes, movies, or songs, though, are the type of words that should work best for use in actual conversation later on.

 Your mini-study is a quick way to get a small sample of the same kinds of data that real-world cognitive psychologists use in their research to learn more about thought processes.

If there are some words that many of your friends give in response to a word, you can assume it is strongly associated with the test word. You want words with the highest probability of priming the mental pump toward a predictable outcome.

Why It Works

The human brain is so efficient that it processes words or ideas in the context of whatever words or concepts have been previously over-learned. Research studies have found that when people are asked to state whether a series of letters is a word, they respond more quickly to words that have been primed or preactivated by words that were shown to them just prior to the identification task. For example, if *stripes* is shown, and then either *tiger* or *lemon*, people will respond more quickly for *tiger* than for *lemon*.

By talking about words or topics that are closely related to other words or topics, you begin a thought process in your friend's brain in which activation of neurons spreads to neurons that generally fire at the same time. Your brain has learned that certain words and topics almost always occur together, so it knows that when one of the associated words or topics is activated, it should also fire in the regions where those associated words and topics are activated. That way, your thought process can proceed smoothly.

Where Else It Works

This particular mind trick has some risk of failure, especially when the associations that you are relying on are low-probability associations. However, you might just enjoy knowing that you are secretly manipulating others and don't have to make a big show out of it.

We can prime people to do lots of things that seem to just come naturally because they occur so effortlessly and often. For example, it is likely that you can make someone yawn simply by yawning yourself. You might even be able to get a friend to yawn by talking about yawning or sleep. (In fact, as I wrote this, I yawned.) Likewise, if there is something that sounds good to you for dinner, you might be able to get your family members to crave it too by mentioning that kind of food.

You probably have been primed yourself many times. When you are listening to your favorite CD and one song ends, do you start hearing the next song in your head before it even begins? If you know what things someone associates with other things, it becomes relatively easy to predict someone's thoughts after you've primed them. This is partially why married people can often finish each other's sentences.

Where It Doesn't Work

If someone doesn't share your language background, because they speak either a different language or a different dialect, they might not have the same word associations that you have.

It also might not work if a word has several equally likely word associations. For example, if you prime someone with the word *hot*, some people might start thinking about the weather (hot and *cold*). Some might think about food (hot *dogs*). Others might start thinking about people they admire (a hot *babe*).

What do you think of next when you see the word *hot*? I *knew* you were going to say that!

—*Jill Lohmeier with Bruce Frey*

Search for ESP

#68

Though most scientists agree that there isn't much evidence that ESP actually exists, they might be wrong. You or your friend or your monkey might have ESP, and there's no time like the present to find out!

The term *extra-sensory perception* (ESP) was coined to describe perceptions that are independent of the traditional five senses: sight, hearing, touch, taste, and smell. The first to use the term was a psychologist at Duke University in the 1920s and 1930s named J.B. Rhine. There was much excitement at the time, as Rhine and his colleagues were able to identify individuals who seemed to exhibit ESP abilities. In the popular press and some of the scientific writing of that period through the 1970s, it was even taken for granted that there was such a thing as ESP and that we all had the trait to a certain degree.

Today, though, you don't really hear much about ESP, and most scientists have concluded that such a thing probably does not exist. More specifically, it hasn't met the criteria for scientific acceptance that any other hypothesized phenomenon is expected to meet, such as experimental evidence, replicated studies, and so on. You can add to the data, though, by conducting your own studies and identifying whether you or your friend might be psychic.

Identifying Psychic Abilities

Though there is a wide range of supposed psychic abilities, ranging from reading minds to moving objects with one's mind, the traditional way to study ESP has been to use a deck of cards called *Zener* cards. Zener decks have 25 cards with matching backs. The face of each card displays one of five symbols: a circle, cross, square, star, or wavy lines, as shown in Figure 6-7.

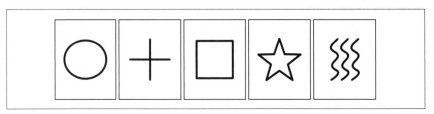

Figure 6-7. Zener cards

If you don't have a deck of these cards handy, you can make them pretty easily with a pack of blank index cards and a black magic marker. Just make sure that no one can see right through them (unless they *are* psychic, in which case they can see right through you, too). Make 5 cards of each symbol for a total of 25 cards.

There are a few different ways you can use a shuffled deck of Zener cards to conduct an ESP test:

- One person tries to guess the order of the cards by announcing each one before it is turned over.

- One person looks at the face of each card and attempts to "send" it to another person telepathically who is sitting nearby.

- A person in another room or in a distant location looks at the face of each card and attempts to send it telepathically to another person over a great distance. Sometimes, the receiver imagines that they are in the room with the sender and can see the card.

With whatever method you choose, the procedure is to go through all 25 cards and keep track of the hits and misses. How many cards out of the 25 did the subject correctly identify? In some studies, the receiver is told how he is doing while they are going through the whole deck; sometimes, he is not told until the end of the experiment. The outcome variable is the number or percentage of cards that were correctly identified.

In ESP research, the person who is trying to read someone's thoughts is the *receiver* and the person who wants her mind read is the *sender*.

Analyzing the Results

If the results are what would be expected by chance alone, treat the outcome as evidence that the subject is not psychic. If the subject gets many more correct than would be expected by chance, that outcome suggests that the subject might have ESP.

So, what would be expected by chance? If you are guessing for 25 cards and there are five cards of each type, chance alone would get about 5 correct. Imagine, for example, that you guessed *star* every single time across all 25 times. You would be guaranteed 5 hits and 20 misses because you know *star* will come up exactly five times overall. If you guessed randomly each time among the 5 possibilities, your average success would also be 5 out of 25, or 20 percent.

What if you had a higher success rate than 20 percent, though? What if you were correct 6 out of 25 times, for a success rate of 24 percent? Should we treat that as evidence that something other than chance is playing a role here? What we need is a statistical analysis of the different possible outcomes, to identify what percentage should be considered so unusual that it must be evidence for the presence of something so unusual.

A statistical test reveals only whether chance is the best explanation for an outcome. For our experiment, a statistically significant outcome doesn't prove that ESP is at work, only that chance is not the best explanation. After all, the best explanation for a high hit rate might be that the receiver sees the cards reflected in the sender's glasses, or some other less interesting cause.

We know that over the short run (or in a *small sample*, to use the stats jargon), results that differ from the population are common. We also know, though, that a large difference from that population value is uncommon, especially over the long run (or with a *large sample*). In fact, the probability of finding a difference of a given size between a sample value and the population value is directly related to the size of the sample.

For ESP experiments, the sample size is the number of guesses or trials, and the population is the known distribution of the different symbols across all trials. The population value for any number of guesses is 20 percent correct; that is what would be expected by chance. If there is a large difference between the sample value and this population value, then something other than chance is likely operating.

The statistical analysis appropriate here is something called the *Z-test for comparison of an observed proportion to an expected proportion*. It is similar to other common statistical tests, such as *t tests* [Hack #17], which calculate a difference and determine how frequently such a difference would be found if a given sample really was randomly drawn from a population with certain characteristics.

The probability of any difference is based on the size of the sample. For example, if after 25 attempts, a person guessed 24 percent correctly instead of the expected 20 percent, the information needed for this analysis would be:

- A sample size of 25
- An observed proportion of .24
- An expected proportion of .20

Without showing the formula and calculations for this particular analysis, I'll show you the result. By chance alone, with 25 guesses, a subject will guess at least 24 percent of the cards correctly 31 percent of the time. Another way of saying that is that out of 100 subjects going through your study, 31 of them will get this result or better. So, a hit rate of 24 percent is better than average, but not so unusual that I would call *The National Enquirer* just yet.

What about other hit rates or if you test with more than 25 trials? Table 6-17 shows the chance of guessing given percentages of cards (or higher) correctly. This table assumes an expected hit rate of 20 percent.

Table 6-17. Likelihood of selected ESP hit rates

Number of guesses	Percent correct (hit rate)	Probability of hit rate or better
25	20 percent	50 percent
25	30 percent	11 percent
25	40 percent	1 percent
25	50 percent	.01 percent
100	20 percent	50 percent

Number of guesses	Percent correct (hit rate)	Probability of hit rate or better
100	30 percent	1 percent
100	40 percent	.00001 percent
100	50 percent	.000000000001 percent

Notice the dramatic drop in likelihood for extreme outcomes as the sample size increases. For example, with just 25 guesses, the chances of getting 40 percent correct is about 1 percent; if you went through a pack of 25 cards 100 times, you are likely to do that well or better just one time. If you took 100 guesses, though, maybe going through the deck four times, you would get 40 percent or better correct just 1 out of 100,000,000,000,000 times!

How Much Is Enough?

If you want to conduct ESP experiments, you should establish a standard for how unlikely a performance must be for you to consider it evidence that something other than chance is the operating factor. Typically, in statistical research, if a result is likely to occur by chance 5 percent of the time or less, the result is considered statistically significant. For ESP experiments with 25 Zener cards and 25 guesses, you will guess 8 or more cards correctly about 7 percent of the time. You will guess 9 or more correctly just 2 percent of the time. So, some standard between 8 or 9 hits is scientifically reasonable.

The skeptic in me feels compelled to leave you with a warning. If you perform this experiment and get a significant result on yourself or someone else, that's pretty cool. If you can repeat the finding, though, replicating the experiment with the same person and getting similar results, that's when it will start to get exciting! If that happens, send me a telegram immediately. I'll sell my house, buy a train, and we'll hit the road to fame and fortune!

HACK #69   Cure Conjunctionitus

The probability of two independent events both happening can never be more likely than either of the events happening alone. Surprisingly, this common sense truth is not commonly sensed.

Imagine that you are introduced to John, a tall, pleasant, athletic-looking man at a dinner party. You chat with John for a few minutes and discover that he is friendly and quick to laugh, but not exactly bright. John is eager to talk about the currently ongoing World Series and also asks you about the car you drive.

On your way home from the dinner party, your spouse asks you about the man you were chatting with before dinner. You share a little bit about John, but realize that you never learned what he does for a living. In fact, as you realize, you really don't know that much about him. Your spouse decides to play a little mind game with you and explains:

> I know a little about John. I'm going to provide a series of statements about him. They might be true or not true. All might be true. All might be untrue. There might be a mix. I want you to place the statements in order based on how confident you are that each statement is true. When we are done, I'm going to diagnose whether you suffer from a common brain ailment known as *Conjunctionitus*.

Your spouse then asks you to rank the following statements, guessing which are most likely true about John:

1. John is a computer scientist.
2. John is a car salesman.
3. John is a former baseball player.
4. John is a Republican.
5. John is a computer scientist who used to play baseball.
6. John is a preacher who runs marathons.
7. John plays the clarinet.
8. John is married.

You, like many other people, might have ranked statement 3 (former baseball player) as one of the most likely possibilities and 1 (computer scientist) as one of the least likely. So far, not so crazy; at least they are reasonable guesses based on the conversation you had.

The symptom related to Conjunctionitus has to do with the position you assigned statement 5 in your rankings. I'm betting you ranked it as more likely than 1. If so, you might suffer from Conjunctionitus, a condition that results in people making poor probability judgments.

The truth is that the probability of two events occurring together can never be greater than the probability of either one occurring alone. Thus, it cannot be more likely that John is a computer scientist who used to play baseball than it is that John is a computer scientist. Never fear, though; the first step in improving your ability to make likelihood judgments in these situations is to admit you have a problem. The next step is to understand the condition, so that healing can begin.

The Problem

Although more information might make a description seem more similar or representative of someone or something, more information does not make something more likely. As mentioned earlier, the probability of two events occurring together cannot be more likely than one of them occurring alone. Consider all of the possible things a man can be in the world. How do you decide which things John is most likely to be? You could start by looking at *base rates*.

There are probably more married men in the world than there are computer scientists, car salesman, former baseball players, Republicans, preachers, marathon runners, and clarinet players. Thus, it is most likely that John is married. Where did you rank that possibility?

Because we probably don't really know the base rates of all the other possibilities, we can use the information we have about John to predict which of the other possibilities is most likely. We do know that if we consider the group of all former baseball players and the group of all computer scientists, there will probably only be a small number of men who belong to both groups. Thus, the likelihood of being in that group of computer scientists who used to play baseball must be smaller than the likelihood of being in the group of computer scientists or of being in the group of former baseball players.

Most people, however, even though they are rational, intelligent decision makers, will be drawn toward sentences that are *conjunctions* (i.e., that list two separate "facts"), as if the listing of the "facts" together makes them more likely to be true. Even if, and maybe especially if, the second "fact" by itself seems unlikely.

Conjunction Junction, What's Your Function?

Why do our minds tend to work this way? In the 1970s, Nobel Prize winner Daniel Kahneman and his colleague Amos Tversky presented college students with several problems in which one option was highly representative of a given personality description, one option was incongruent with the description, and one option included both the highly similar and the incongruent options.

Perhaps the most well-known problem that demonstrates the conjunction fallacy is the now-famous (at least in cognitive psychology circles) Linda Problem:

> Linda is 31 years old, single, outspoken, and very bright. She majored in Philosophy. As a student, she was deeply concerned with issues of discrimination and social justice, and she also participated in antinuclear demonstrations.

Subjects were asked to rank these statements based on high likely they were to be true:

1. Linda is a teacher in elementary school.
2. Linda works in a bookstore and takes Yoga classes.
3. Linda is active in the feminist movement.
4. Linda is a psychiatric social worker.
5. Linda is a member of the League of Women Voters.
6. Linda is a bank teller.
7. Linda is an insurance salesperson.
8. Linda is a bank teller and is active in the feminist movement.

Kahneman and Tversky (and many others who have since replicated their work) found that people consistently ranked option 8 (a bank teller active in the feminist movement) as being more likely than option 6 (a bank teller). This is because option 8 provides more information, which seems to be more representative of Linda. Because we expect her to be politically active, but we don't expect her to be a bank teller, it seems as though the only way she could be a bank teller is if she is also politically active.

However, we know that 8 can never be more likely than options 3 or 6, because if we imagine all people active in the feminist movement, a subset of them (perhaps a small subset) will be bank tellers. Likewise, if we imagine all of the bank tellers in the world, a subset (again, perhaps a small one) will be active in the feminist movement. Thus, the likelihood of being a bank teller must be greater than the likelihood of being a bank teller who is active in the feminist movement. Makes sense, right? But your mind doesn't want to work that way.

> The rule that states that the probability of two events occurring together cannot be greater than the probability of either one of them occurring alone is called the *conjunction rule*. The fact that many people often believe that the conjunction of two events is sometimes more likely than one event occurring alone is called the *conjunction fallacy*.

The Cure

To stop thinking wrongly about these sorts of propositions, the cure is simple:

1. Cut it out.
2. Stop.
3. Don't do that.

The conjunction fallacy can be seen at work in numerous places. Be aware of situations in which it might occur and analyze the situation. For example, you can ask a baseball fan about a favorite player who doesn't often hit home runs. Ask whether the player is more likely in the next game to do which of the following:

- Hit a home run
- Strike out
- Strike out *and* hit a home run

The fan probably believes that a home run with a strikeout in the game is more likely than just a home run. But it cannot be so.

There are some situations in which it might be okay to pick the conjunction proposition. If two things *must always* occur together (such as thunder and lightning), then the likelihood of both of them occurring is the same as one of them occurring. And if you add to the thunder and lightning statement and change it to the likelihood of thunder (and no lightning) versus the likelihood of thunder and lighting, then, in fact, the likelihood of thunder and lightning is more probable. However, this is true only if one can *never* occur without the other.

Once you are aware of this common error in probability estimation, you can see it everywhere. For example, one place in which you can readily find the conjunction fallacy is in the political prediction arena. Is George W. Bush more likely to:

- Nominate a moderate Supreme Court justice
- Nominate one moderate Supreme Court justice *and* one right-wing Supreme Court justice

Of course, you know the answer now, but many political analysts might argue with you. But that's because they have the sickness. They have Conjunctionitus. You did too, once, but now you are cured.

See Also

- Tversky, A. (1977). "Features of similarity." *Psychological Review*, 84, 327–352.
- Tversky, A. and Kahneman, D. (1974). "Judgment under uncertainty: Heuristics and biases." *Science*, 185, 1124–1131.

—*Jill Lohmeier*

Break Codes with Etaoin Shrdlu

HACK #70

You never know when you will have to decipher a cryptic message, whether it's one intercepted by your man, James Bond, or one scribbled illegibly onto a prescription pad by your doctor. Here are all the statistical tricks you'll need, Agent 003.14159.

You might have noticed that certain keys on your computer keyboard get dirty or wear out more quickly than others. That's because you hit them more often than the others. You might also notice that these letters tend to be in the middle of the keyboard or, more correctly, in small circles near where your hands are when they are centered on a keyboard.

Both the wear and tear on your keys and the placement of them in a standard typewriter (a.k.a. QWERTY, for the first six letters on the top row) pattern are based on their frequency of use in English. Different letters in the alphabet are used with different frequencies in the spelling of words in a language. By applying the known frequency of these letters, along with other statistical tricks, you can quickly decode classified documents, whether they are Leonardo da Vinci's diary, a puzzle in the newspaper, or big, bright letters being turned by Vanna White on TV.

Single Substitution Ciphers

The simplest and oldest type of letter-based code is the *single substitution* format. In these codes, some message is transformed from the actual letters in the words to other letters in the alphabet. In the simplest form of this type of coding, the same letter substitutes for the same letter throughout the message. For example, a simple cipher might use the substitution pattern shown in Table 6-18, in which the letters on the top row (the *plain text*) are replaced by the letters on the bottom row (the *cipher text*).

Table 6-18. A single substitution cipher

Plain text	A	B	C	D	E	F	G	H	I	J	K	L	M	N	O	P	Q	R	S	T	U	V	W	X	Y	Z
Cipher text	N	A	O	B	P	C	Q	D	R	E	S	F	T	G	U	H	V	I	W	J	X	K	Y	L	Z	M

With a code like the one shown in Table 6-18, the following plain-text passage:

> Tom appeared on the sidewalk with a bucket of whitewash and a long-handled brush.

appears in cipher text like this:

> Jut nhhpnipb ug jdp wrbpynfs yrjd n axospj uc ydrjp yhwd ngb u fugq-dngbfpb aixwd.

The passage looks like nonsense, but with the key shown in Table 6-18, anyone could easily replace the nonsense letters with the original letters, causing the opening sentence of the second paragraph in Chapter Two of *Tom Sawyer* to reveal itself.

Using Probability to Decode Substitution Ciphers

Of course, the real task when deciphering ciphers is to do it without access to the code key. Real-life code breakers and winning contestants on *Wheel of Fortune* use the same tool to solve their problems: they apply the known distribution of letters in English language words.

The advent of computers, computer analysis, and electronic copies of millions of books has made the calculation of exact probabilities for each letter of the alphabet possible, though cryptographers (code makers and breakers) have known the basics for some time. Here are some of these basics:

- The most common letter, in terms of usage in English, is E.
- The least commonly used letter is Z.
- The most common consonant is T.
- J and X are rarely used, as is Q.
- When Q is used, it is almost always followed by U.
- Only A and I are used as one-letter words in English.

With even just these basic probability facts, you could begin to tackle decoding a cipher such as our Mark Twain passage. The most commonly appearing letters in the garbled version are P and N. Because N is used as a single-letter word, it cannot be E (N is most likely A), so a good first guess for P is that it substitutes for E.

With just a little knowledge of letter distribution, we have already identified the substitutes for E and A. We can't be sure we are right, but like any good statistician, we think we are probably right. Table 6-19 shows the likely distribution for each letter of the alphabet.

Table 6-19. Frequency distribution of letters in English

Letter	Frequency
A	8.04 percent
B	1.54 percent

Table 6-19. Frequency distribution of letters in English (continued)

Letter	Frequency
C	3.06 percent
D	3.99 percent
E	12.51 percent
F	2.30 percent
G	1.96 percent
H	5.49 percent
I	7.26 percent
J	0.16 percent
K	0.67 percent
L	4.14 percent
M	2.53 percent
N	7.09 percent
O	7.60 percent
P	2.00 percent
Q	0.11 percent
R	6.12 percent
S	6.54 percent
T	9.25 percent
U	2.71 percent
V	0.99 percent
W	1.92 percent
X	0.19 percent
Y	1.73 percent
Z	0.09 percent

ETAOIN SHRDLU

The strange phrase "ETAOIN SHRDLU" is a *mnemonic device* (memory tool) for remembering the most frequently occurring letters. These 12 letters account for over 80 percent of total letter frequency.

You might notice that the order of letters in ETAOIN SHRDLU is not exactly the rank order of popularity shown in Table 6-19. It is close enough, though, and easier to pronounce than if it were exactly correct. Another thing to remember is that any "definitive" list of letter probability depends on the source material for the letter count. You can find many different lists of letter order and frequency, and some differ slightly from others.

For example, one organization that produced a list of statistical distributions of letters in English text relied on a computer analysis and actual count of letter occurrence in seven literary classics, such as *Jane Eyre* and *Wuthering Heights*. Two of these seven books were Tarzan novels. I'm guessing that if we were to compare that table of letter distributions with others, we would find that the proportional number of times the letter Z appeared was greater than if other sources were used. For the common letters, though—such as E, T, and A—there is wide agreement on their use as best first guesses for code breaking.

Wheel of Fortune Strategy

On the TV game show *Wheel of Fortune*, before solving the big puzzle at the end, the producers are nice enough to provide certain letters and show whether they appear in the hangman-type phrase. They provide R, S, T, L, N, and E. These are given, of course, because they are common letters, and are in our top 12: ETAOIN SHRDLU. The player is allowed to choose three more consonants and another vowel. Using our statistical knowledge of letter frequency, a good basic strategy would be to pick A as the vowel and the three most common consonants not yet shown: H, D, and C.

Statistical Analysis of Coded Texts

Here's how you might use these letter stats in real life to decode a secret message or solve a puzzle. This method works best if the coded text is lengthy, but it works surprisingly well even for shorter passages. Calculate the distribution of the coded, substitute letters (the cipher text), and then compare it to the distribution shown in Table 6-19.

Figure 6-8 shows how this process might look graphically. Only the first 10 most common letters are shown, but the analysis would use all the letters. This example pretends that there is a lot of coded text and that the substitute cipher shown in Table 6-18 is being used.

Because the most common substitute letters are P, followed by J, a good guess for breaking the code would be to see whether P could really be E and J could really be T. These first guesses can be made all the way down the line for each letter. By starting with the most frequently appearing letters and moving down the list, a code breaker can quickly see whether these first hypotheses are right or wrong and change guesses around until English words start to appear.

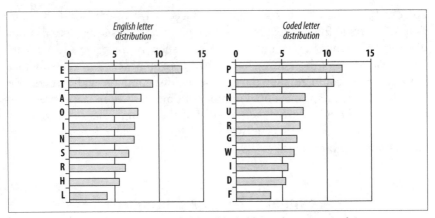

Figure 6-8. English letter frequency (left) and coded letter frequency (right)

Other Common Letter Patterns

Beyond just knowing the frequency of individual letters appearing, good code breakers use probability information about other patterns of letters:

- Words are most likely to start with T, O, A, W, or B.

- Most words end with an E, T, D, or S.

- If two letters are doubled in a word, they are most likely to be SS, EE, TT, FF, or LL.

- Frequently appearing two-letter words include *of*, *to*, *in*, *it*, and *is*.

- By far, the most common three-letter words are *the* and *and*. Other common three-letter words include *for, are*, and *but*.

- Letters that tend to come in pairs include TH, HE, AN, IN, and ER.

- The most frequently used words are *the, of, and, to, in, a, is, that, be*, and *it*.

- Perhaps indicating what people tend to write about, the top 100 most-used words in written texts include *dollars, great, general*, and *public*. *Debts* just barely failed to make the top 100, but it is surprisingly common.

See Also

- A good explanation of single substitution ciphers can be found under the entry for *frequency analysis* at *http://en.wikipedia.org/wiki/Frequency_analysis*.

- Some of the statistics reported in this hack were found at *http://www.data-compression.com* and *http://www.scottbryce.com*. Good information and advice for solving cryptograms and other codes using statistics can be found at those sites.

Discover a New Species

While everyday entire species of creatures become extinct, occasionally new species are identified that were previously unknown. Surprisingly, statistical tools, not biological tools, can do the trick.

A few years back, a new species, a type of possum, was identified. The new species was named *trichosurus cunninghamii*. *Trichosurus* means, um...possum (I guess), and the *cunninghamii* part refers to its discoverer, Ross Cunningham, a statistician at Australian National University. If you'd like to have a species named for you, here's how statistics can help.

Identifying Species with Statistics

There is a family of statistical analyses that looks at a bunch of variables and finds naturally occurring groupings among them. Typically, the groupings or clusters of variables are identified on the basis of the correlations among them [Hack #11].

One procedure that uses this strategy attempts to find underlying dimensions or invisible, giant basic variables that account for a bunch of less important variables. This procedure is *factor analysis*, and elsewhere we see how it can, among other things, be used to identify writers' styles [Hack #65].

Statistics is full of similar techniques that can identify dimensions, underlying causes, and groupings. The goal of identifying groupings is of greatest use to biologically inclined statisticians who wish to identify new species.

For some group of animals to technically be a separate species, it must share a unique set of biological characteristics that make it distinct from similar animals. Sure, animals within the same family all look a little different from each other, but then, people look a lot different from each other and we are all one species (my Uncle Frank being perhaps the exception that proves the rule).

If a group of animals, such as Dr. Cunningham's possums, have more in common with each other than they do with the other creatures in their species, they might be candidates for consideration as a species in their own right. Statistics can determine that "more like each other and more different from the rest of the species than chance alone would produce" point.

Using Cunningham's discovery as a model, there are a few steps to follow for you to make your own discovery.

Collect some data. This possum existed in Australia near people for more than 200 years and no one noticed. To be fair, it looked an awful lot like the other possums, the most common of which was the *trichosurus caninus*, now called the *short-eared possum*.

It was assumed for some time that there was really just this one species of the little guys. Part of Dr. Cunningham's job was to collect and organize descriptive data for the wildlife around him. Consequently, he had a ton of very specific quantitative descriptions of various possum parts—eyes, ears, nose, and throat—and measurements of other physical characteristics.

Choose a statistical method. Cunningham's choice was a technique similar to factor analysis but with a more imposing name: *canonical variate analysis*. You can use any method that uses the variability in scores to create distinct groupings. Some of those are discussed in this book—such as factor analysis, mentioned earlier in this hack—but there are many other procedures that would work.

> If you are really statistically savvy, it will help you to know that *canonical variate analysis* is functionally the same as *discriminant analysis* or *multivariate analysis of variance* (MANOVA), two other procedures that create linear composites of variables with the goal of conceptually defining two or more distinctly different groups.

Cunningham used this statistical procedure to examine the descriptive data for this presumably single species (you know, these *trichosurus caninus* fellers) and demonstrated that there were likely two different species.

Select a hypothesis and analyze the data. Statisticians test *hypotheses*, so you should begin your analysis with a guess about whether there *is* or is *not* a distinction between the groups of participates who supplied your data.

In the example of our hero, Cunningham assumed that there were two different groups of critters that accounted for the data. Then, the procedure (using a computer for the calculations, of course) identified which variables worked best as key distinguishing characteristics between the theoretical groups.

> The difference between using this tool, canonical variate analysis, and something like regression is that, when using variables to make predictions in regression, the researcher has some known data about scores of actual subjects: which "group" they belong to [Hack #13]. Here, the procedure works blindly without knowing what the correct answer is. Instead, it finds groups that can be made the most different with the variables at hand.

Here are the variables Cunningham used:

- Head length
- Skull width
- Eye size
- Ear length
- Body length (from tip of nose to tip of uncurled tail)
- Tail length
- Chest width
- Foot length

While other variables were considered, Cunningham chose these because they were eventually found to be most important in distinguishing one species from another and also because they were characteristics that would probably be unaffected by environment.

Interpret results. The last step in any statistical analysis is to describe and understand whatever you found. For discovering species, you need to be able to describe that new species in enough detail to differentiate it form other, similar species.

The procedures used by Cunningham identified a series of different equations that weighted each of the biological variables differently, to find the combination that best identified two separate groups. These equations (which the procedure labels *variates*) are similar to regression equations, with the outcome or criterion variable determining which group a possum belongs to.

Here's the single best equation that accounted for an astonishing 89 percent of the variability on these characteristics for all the possums in his database:

(head length $\times$ .44) + (skull width $\times$ .07) + (eye size $\times$ .05) + (ear length $\times$ .82) + (body length $\times$ .35) + (tail length $\times$ .72) + (chest width $\times$ .16) + (foot length $\times$ .70)

I've provided the standardized weights from the study, so we can compare them to each other. The larger weights indicate the possum parts that differed the most between the mathematically chosen two groups of possums.

In this data, you could find two groups of possums that differed the most based on ear length, tail length, and foot length. The amount of variability explained was so large that, statistically, Cunningham concluded that the mathematically identified groupings were real. The two groups of possums found in the data were actually two different species of possum, and the species could be defined by their ear length and a couple of other variables. The larger the weights in the equation shown earlier, the more the two species differed on these body parts.

Two Possum Species

Table 6-20 shows the official descriptions of the two possum species first identified as such by our statistician and his mathematics. Notice the names are even based on the key predictors found in the statistical analysis!

Table 6-20. Two common Australian possums

	trichosurus caninus	trichosurus cunninghamii
Common name	Short-eared possum	Mountain brushtail possum
Habitat	Lives in the north	Lives in the south
Ears	Shorter ears	Longer ears
Feet	Smaller feet	Larger feet
Head	Bigger head	Smaller head
Tail	Longer tail	Smaller tail

So, start collecting your own data on those odd, stinky bugs you find on your screen door and you are well on your way to greatness and immortality. Is there one species of stink bug or two? You tell me.

See Also

- I first learned about this approach to identifying species in this fine article: Hall, P. (2003). *Chance, 16*, 1.

Feel Connected

HACK #72

The concept of "six degrees of separation" is more than just a New Age metaphor for community or a party game involving the actor Kevin Bacon. If you want to actually test the idea that we all know someone who knows everybody else, find out how closely linked you really are to everyone.

I know a guy who knew a guy who used to work for the President of the United States. Small world, eh? I'm not saying I have great connections, but I am just two handshakes away from the leader of the free world. Before you get too impressed, you should know that you probably are just a few links away from almost anybody in the world.

It is probably true that any two people are within *six degrees of separation*, and that magic and oft-quoted number of 6 is actually taken from a real scientific study! Here are some clever research methods to let you reveal the invisible connections that unite us all, or at least link you to that person on the other side of the cocktail party.

Six Degrees of Separation

There is a play called *Six Degrees of Separation* by John Guare and a movie based on that play starring Will Smith. There is also a popular party trivia game, sometimes called *Six Degrees of Kevin Bacon*, that attempts to link any actor or actress through a series of movies and other performers until they share a connection with actor Kevin Bacon.

The phrase and concept come from a study that considered the *small-world problem*. Have you ever been at a party or been chatting with a stranger at a coffee shop and discovered that you both know the same person? Social psychologist Stanley Milgram was curious about this phenomenon in the late 1960s (when there were a lot more cocktail parties than there are now). How much overlap was there in social networks? If we could all get together and list everyone we know, would there always be some connection? Probably, eventually, as we explored further and further out of the center of our web of acquaintances, we would find some connection with almost everyone. But how many links would it take?

Just one degree of separation means we all know everyone. Well, I don't know you (no offense), so we know that one is too few links to connect everyone. Are there just two degrees of separation? If we don't know each other, maybe we have a friend in common?

The question, therefore, is how many degrees of separation *are* there between you and anyone else? To get the answer, do a big study or a small study using the methods in this hack.

Doing a Big Study

How could one study the problem of whether we actually live in a small world? The best way is to duplicate the methods used by Stanley Milgram.

Choose a target. Milgram started by picking someone he knew who worked in Boston, Massachusetts, where Milgram lived. It wasn't Kevin Bacon, but a stockbroker who agreed to act as the target, the final end of a chain that Milgram hoped to build. You could pick your best friend or your school principal or your University's president. You gotta ask their permission first, though (something about ethics).

Recruit participates. Milgram then randomly sampled from two communities: Boston and Omaha, Nebraska. This sampling scheme was meant to represent the two extremes of likelihood that anyone would know the target. Start with people close by and people far away, and the average of their data should be fairly representative of the population. Milgram used 300 randomly chosen recruits. You should use as many as you can afford or have time for.

Train participates. Milgram sent a packet in the mail to each recruit. The packet contained instructions describing the study and a letter for the Boston broker. They were asked to deliver the letter to our guy, but only if they knew him personally. If they did not know him personally, they were asked to record some information, such as their name, and send the packet on to someone who they did know who they thought might have a better chance of knowing him. Those next people in the chain received the same packet with the instructions and the letter. They might have sent it to the broker if they knew him, or sent it on to a third link in the chain, and so on.

In your own study, make sure to write the instructions clearly and simply, and, these days, you might explain that this is legitimate research, not a commercial solicitation and not a chain letter (though it literally is, I guess), and all the disclaimers you think will help. You should also include contact information for you if anyone has any questions about the legitimacy of the project.

Collect and analyze the results. After a reasonable amount of time, check with your target and gather all the letters received. On each letter, count the number of names that form the chain. Average all the different lengths of chains to determine the typical number of connections. Find the smallest number necessary to include even the longest chain, and you have the maximum distance.

The Boston target in Milgram's study eventually received about 100 letters. Of those, the average number of links was six—thus, the origin of the number six in "six degrees of separation."

Notice, however, that not all letters arrived, so we don't know from this one study that six is really the right number. The study also took place in the U.S. only, not worldwide, so grander views of there being only a few degrees of separation between any two people on the whole planet are philosophically based, not empirically derived.

> The *response rate* that Milgram enjoyed was very high, considering the complicated requests made of participants. This is not surprising, because Milgram knew something about obedience. Stanley Milgram is probably better known for another clever study with more disturbing results he conducted some years before his *small world* study. With his obedience studies of the early 1960s, Milgram demonstrated that when people of authority (such as research assistants in lab coats) ask study participants to do something that makes them uncomfortable, such as administering (or believing that they are administering) an electric shock to another research subject, a surprising number of people will do it. His research led to much insight as to why people might "obey orders" even if they disagree with them.

Two more recent studies have confirmed that the average number of connections between people in social networks is about six or even a little less.

Doing a Small Study

There are a couple of ways to use these methods that don't take quite as much work. The goal of the activity could be scientific or just party fun.

Milgram via email. Duplicate the Milgram study, but use the convenience of email. Here, the question would be how many links between people using their email addresses. Email is easier to work with than snail mail and is virtually cost-free.

Of course, choosing recruits through email is probably more difficult. It is hard to choose email addresses randomly, because there isn't a big phone-book-type list to sample from. Also, your email requests might quickly be mistaken(?) for spam and ignored. By the way, because your research interest is legitimate, you shouldn't have to worry about violating any Internet protocols.

Throw a party. When hosting a large party (Milgram would have loved it if you used a cocktail party, the inspiration for his original study), hand out supplies to your guests. Give them each a large index card and a pen. At the bottom of each card, list the name of a guest at the party. If guests don't know the person listed below, they should sign their name at the top of the card and hand it to someone else who they think might know the person.

The process should continue, just as in the Milgram study, until the cards reach the person who is named on the bottom. That person then turns the card in. At the end of the party, you can analyze the data and prove to your guests that they all really know each other.

Just Doing the Math

Even without scientific studies, however, a quick mathematical analysis might convince you that the number of people between you and anyone else is a fairly low number. How many people do you know by their first names? 100? 200? Let's say it is about 100. They each know about 100 people by their first names, too, presumably, so you are already connected to 10,000 people through just two degrees of separation. (Actually, 10,100, in total, counting the 100 people who are within one degree of you.) It wouldn't take too many degrees before you are connected to a whole lot of people, as shown in Table 6-21.

Table 6-21. Degrees of separation and corresponding connections

Degrees of separation	Connections
1	100
2	10,000
3	1,000,000
4	100,000,000
5	10,000,000,000

In fact, with just five degrees of separation, you should be connected to 10 billion people, more than there are on earth!

So, why, in reality, are a greater number of connections needed to really link all people? The problem is that the groups of 100 people that each person knows are not independent of each other. There is not a different group of 100 friends for each of your 100 friends. A good proportion of the 100 people that you know well are on many different lists for that group.

There is much overlap in social networks. This overlap actually helps increase the chance that there will be a fairly direct link between you and anyone else who lives relatively close to you (in the same country, say).

The Grandparent Paradox

A similar problem related to network overlap is the *Grandparent Paradox*. You had two parents. Your parents had two parents each, so that's four grandparents. Each grandparent had two parents and four grandparents. You don't have to count back more than a few generations to get to a huge number of people.

Count the grandparents going back 40 generations ago, and you require a trillion people. That's more than have ever lived on earth for all time combined. And that's for just the last 1,000 years or so. Where did we get all these other grandparents? Jupiter, perhaps?

The answer, of course, is that somewhere along the way there must have been some overlap on the genetic tree. Some already related family members must marry occasionally and have children. For the sake of decorum, I'll suggest they were second cousins or something like that.

The technique that Milgram used—the *small-world method*—has been found to be very useful in all sorts of social network research. The concept of a few degrees of separation has an intuitive appeal because it makes us all feel part of a small community.

It is also reinforced every time we do find a connection with a stranger through some common friend. I don't know about you, but in my own world, I have such importance that I can easily connect myself to all sorts of famous people. For example, as a college student at the University of Kansas in Lawrence, Kansas in the early 1980s, I was an extra in the ABC TV-movie *The Day After*, an acclaimed film about the potential after effects of nuclear war in the U.S. *The Day After* featured actor John Lithgow as a science professor. Lithgow later appeared in the film *Footloose*, starring Mr. Kevin Bacon! It's a small world, after all.

See Also

- The two more recent studies confirming six or less degrees of separation are described in the *Psychology Today* article "Six Degrees of Separation" by Darby Saxbe, which appeared in the November/December 2003 issue.

- Watts, D.J. (2003). *Six degrees.* New York: Norton. A book on the new science of networks provides a comprehensive and fascinating discussion of the connected age in which we live, including the six degrees of separation concept:

Learn to Ride a Votercycle

#73

Though a free election seems to be the fairest and wisest system for making policy decisions and electing officials, statisticians sometimes fear that a paradox political scientists call "vote cycling" can result in a win for the minority. There's a better way to hold an election.

When I was a little child statistician, my parents would occasionally allow me to make choices about personal things—what to wear, what to eat, which story book to read at bedtime, and so on. I noticed that sometimes the choice was open-ended: "Your choice, Bruce: when would you like to go to bed?" And sometimes the choice was presented as a set of alternatives for me to choose between: "Your choice, Bruce: would you like to go to bed *now* or in *five minutes*?"

Of course, the second choice isn't much of a choice, really. When I had to choose between various alternatives, my true opinion wasn't reflected as accurately as when I could choose anything I wanted.

Democracy works like that as well. When it is time to vote for President, or Mayor, or Dogcatcher, we usually must choose between several alternatives. We might not be happy with any of the options, but we vote anyway (at least statisticians do).

Did you ever leave the voting booth, though, and feel that somehow your real feelings weren't represented very well by those choices? Political scientists know that feeling. They have analyzed the sometimes unsatisfying outcomes of votes between alternatives and discovered that such a process can result in outcomes in which no one is happy (except the winner, of course).

Vote Cycling

There are a variety of ways that elections can be structured. Imagine that an electorate (such as the residents of a city, members of a club, or faculty at a university) is asked to vote on a policy and there are three choices. Imagine, also, that there are three groups of supporters that each favor one of the three options over the others. The election could ask people to vote for their favorite policy. Under that system, the policy favored by the largest group is likely to win the most votes. This seems fair, and this is the system we most commonly see.

Another system that makes good sense, too, at least on the surface, would be to present each pair of options against each other and have a kind of *run-off* election, in which A is compared to B, B is compared to C, and C is compared to A. The biggest vote getter in this sort of system should result in an equally fair decision. It turns out, though, that this type of system, called *vote cycling*, is difficult to use fairly because the order in which you present the options can determine the outcome of the election!

Vote cycling in elections works in the same way as how you put together a basketball tournament: the order in which the games occurred could affect who wins the whole thing.

How It Works

Here's an example of how vote cycling can work. Imagine that your scout troop has to decide what color to paint the inside of the troop clubhouse (or wherever scouts meet these days). As a group, you will be voting for Red, White, or Blue. Different political "groups" have formed among your colleagues who favor different color choices.

There are the *Apples* who prefer red, the *Elephants* who favor white, and the *Jayhawks* who like blue best. The groups also differ on which color they like second best and which color they like least. Table 6-22 shows the three groups and their political agendas.

Table 6-22. Painting preference and politics

Group	Percentage of electorate	First choice	Second choice	Third choice
Apples	20 percent	Red	White	Blue
Elephants	40 percent	White	Blue	Red
Jayhawks	40 percent	Blue	Red	White

To determine the will of the scouts, you could hold a two-stage election. Stage one presents two alternatives. The winner of that stage then "competes" with the third alternative to pick a winner. The two stages and results could look like this:

1. Red or White? Referring to Table 6-22, it is likely that Red would receive 60 percent of the vote, knocking out White. Now, the winner goes up against Blue.

2. Red or Blue? In this matchup, Red receives 20 percent of the vote and Blue wins with a huge 80 percent.

So, blue paint must be the will of the people! This is a paradoxical outcome, though, because only one group, representing 40 percent of scouts, liked blue best. An equal number liked white best, and another 20 percent hated blue. The order of decision making affected the outcome. Let's do it again in a different order:

1. Red or Blue? Blue wins with 80 percent of the vote.

2. Blue or White? White wins this match with 60 percent of the vote.

We have a different outcome than before, just because of the order of matchups. This is fun; let's do it one more time. Maybe we can arrange for red to win this time:

1. Blue or White? White will get 60 percent of the vote in this battle and survive to face off with Red.

2. White or Red? Red wins this one, with a majority of 60 percent. Well done, Red. Red clearly is the favorite color!

Three potential orders of matchups result in three completely different policy decisions.

Getting Off the Votercycle

If we think of voting systems as measurement systems, this matchup method of making decisions has low validity. There is information that could be gleaned from the voters that is being lost here. However, there are a couple solutions that come to mind to solve the problem of vote cycling.

If the designers of the voting system are interested in the rank-order preferences of voters, voters could be asked to rank-order all candidates. The lowest mean rank wins. This is a fairer method that uses all the information available, but it can lead to choices that no one is really thrilled about.

For example, such a system resulted in my family's infamous decision to go see *Home Alone* as our Christmas Eve movie many years back.

Another solution is to make all candidates available for a single vote, with the majority winning. This is the most common system, but it does have the disadvantage of sometimes choosing candidates that have no majority support.

For elections in which there are many candidates (in some mayoral or governor elections, for example), there is often a runoff election in which the larger number of candidates is whittled down to a smaller number. This doesn't have the weakness of vote cycling, because all alternatives are considered at the same time. It also eliminates the weakness of the single-trip-to-the polls approach because it increases the likelihood of a winning candidate with majority support.

HACK #74 Live Life in the Fast Lane (You're Already In)

By applying the laws of chance, knowledge of human nature, and some facts about highway-driving behavior, you can make wiser lane-changing decisions.

Nothing is more frustrating than being stuck in a traffic jam, especially when the other cars are moving faster than you. While it is tempting to change to a faster lane, it turns out that your judgment might be flawed and the other lane is probably not really any faster than yours.

Deciding to change lanes when you shouldn't is a dangerous proposition. Not only are the majority of car crashes due to driver error, but 300,000 car accidents each year in the U.S. occur specifically while a driver is changing lanes. Of course, if you are in a hurry and the lane next to you is moving more quickly, as long as you do so safely, why shouldn't a smart driver move into the fast lanes of life? After all, as I've patiently explained to court authorities a number of times, a "good" driver isn't necessarily a safer driver; he's just a driver who gets where he wants to go as quickly as possible.

The problem is that recent research involving statistically based computer simulations suggests that drivers will usually judge another lane is moving more quickly than theirs, even if it is actually moving at the same speed!

This misperception, survey research shows, is enough for most drivers to try to change into that other lane.

Skips, Slips, and Epochs

Our perceptual world while on a busy highway or in a traffic jam consists of the big truck in front of us, the cars we see to the right and left of us, and the poor sap stuck behind us. To judge our speed of travel, while we do have a speedometer, the most compelling data tends to be the cars on either side of us. (Are they passing us or are we passing them?)

Traffic researchers call the times when you are passing other cars *skips* and the times when other cars are passing you *slips*. Recent research refers to skips as *passing epochs* and slips as *being-overtaken epochs*. It probably does not surprise you that drivers greatly prefer passing epochs over being-overtaken epochs.

An *epoch* is a period of time. Drivers' lives while driving in heavy traffic are essentially defined by series of epochs of very short duration.

In addition to looking for faster lanes to move into, drivers have another goal, which is to keep their own vehicle moving as quickly as possible, or at least close to their target speed (which might be the speed limit, for example). If there are perceived gaps between themselves and the vehicle in front of them, and they are not currently moving at their target speed, drivers will accelerate to close the gap. It is these bursts of acceleration that account for the skips (periods of passing other cars) and slips (periods of other cars passing them). We are likely to experience more periods of time when we are being passed than periods when we are doing the passing. It is this perceived inequity that can result in drivers concluding that they are in the slow lane, even if both lanes are equally slow.

Imagine two lanes of traffic side by side that are moving at the same average speed. Gaps between cars form *randomly*; more accurately, they form *systematically*, but based on a random starting configuration. Gaps are filled as they form, and when gaps are filled, cars have accelerated.

Average speed for a lane of traffic can be calculated as distance traveled divided by a period of time. So, if cars in two lanes cover 1,000 yards in five minutes, they both have the same average speed of 200 yards per minute, or 6.8 miles per hour.

Drivers on crowded roads occasionally have gaps they seek to close, but they actually spend much more time (relatively speaking) moving slowly or not moving at all. During those times of slow movement, which, of course, take more time, there will occasionally be cars in other lanes filling gaps and passing the drivers in those temporarily slow lanes.

As measured by epochs, for any one driver there will be more time spent being passed than there will be time spent doing the passing. This is because you pass while moving quickly and you are passed when you are moving slowly. Figure 6-9 paints a picture of this perception.

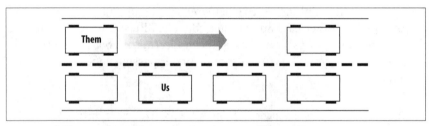

Figure 6-9. The perception of time spent getting passed

Sitting still while watching other cars accelerate to fill gaps creates the illusion that our lane is moving more slowly.

Probability and Traffic Patterns

Canadian researchers Donald Redelmeier and Robert Tobshirani, who conducted computer simulations to determine the accuracy of driver perceptions of other lanes' speed, made some assumptions about traffic patterns that were based on the normal distribution [Hack #23].

To mirror the reality that a particular pattern of spacing on a crowded highway has several causes (conditions, exits and entrances, and so on), they randomly assigned intervals between moving cars based on two normal distributions: 90 percent of intervals were about two meters apart, give or take a 10th of a meter, while 10 percent of the intervals were 100 meters apart, give or take 5 meters. At the start of each of hundreds of simulations, cars were created and spaced following this randomization plan.

The researchers created data for two lanes of traffic moving in the same direction at the same speed, full of hundreds of imaginary vehicles with typical acceleration and braking capabilities. They programmed in a *safe driver* strategy of moving up when there was space in a lane, but not getting too close. Their simulated drivers were not allowed to get too close to another vehicle's tailgate. Also, they were not allowed to change lanes, which must have been frustrating for the little computer-controlled drivers. No accidents here.

> In terms of the average acceleration and braking speed for their simulated vehicles, Redelmeier and Tibshirani chose typical statistical specifications (the ability to go from 0 to 63 miles per hour in 10 seconds and the ability to go from 63 miles per hour to 0 in 5 seconds), which happen to pretty much match a Honda Accord.

Making Wise Lane-Changing Decisions

Redelmeier and Tobshirani found that 13 percent of the time, cars are either passing or being passed. Most of the time, cars are running equal to each other. While there was a better chance that any particular driver was being passed than that she was doing the passing, when she did pass cars, she passed a bunch. The math worked out to a draw in terms of cars passed and the number of cars doing the passing. The total number of cars overtaken by our driver was equal to the number of cars that passed her.

Under crowded driving conditions, the other lane will seem greener much of the time. There are some ways to deal with the misperception and make wiser (and statistically safer) driving choices:

- As a logical scientist, you can evaluate your driving by the length of the journey, not by whether you won or lost the traffic jam competition. It shouldn't really matter if you think more cars passed you than the other way around.

- Keep this *other lane is better* misperception in mind and find better ways to judge the speed of other lanes. Pick a unique car in the other lane, and after a few minutes compare your position to it. After all, there sometimes *are* faster lanes than others; it's just that you can't look at passing cars as the best evidence for speed.

- On large highways, pick a lane far to the left or right of upcoming exits, as traffic exiting and entering the road is the main cause for slow-downs and speed-ups.

- Curb your aggressive tendencies in both driving and in car purchasing. Interestingly, the simulations showed that aggressive driving, such as minimizing the standard following distance between you and another vehicle, will actually increase the amount of time you'll notice other cars passing you. Also, faster cars (those that can accelerate quickly) spend less time passing because they can do it quicker. So, your super-powered sports car might lead to more frustration for you on crowded highways.

The wisest tactic when it comes to dealing with the likely misperception that the other lane is faster than yours might be the simplest. Just don't pay attention to it. The simulations show that if you look at other lanes half as often, you'll notice cars passing you half as often.

I suppose, though, that we don't need a statistical analysis to tell us this. Instead of the cars beside you, pay more attention to the cars behind you. You're way ahead of them and there are thousands of them. You've already won that game.

See Also

- Redelmeier, D.A. and Tibshirani, R.J. (1999). "Why cars in the next lane seem to go faster." *Nature, 401*, 35. The original study reporting this most recent traffic analysis.
- Redelmeier, D.A. and Tibshirani, R.J. (2000). "Are those other drivers really going faster?" *Chance, 13*, 3, 8-14. A more detailed description of the findings reported in the *Nature* article.

HACK #75   Seek Out New Life and New Civilizations

The search for extraterrestrial life is alive and well. You can use statistical sampling and probability to focus the search.

The scientific quest to make contact with life on other worlds requires that decisions be made. First, one must decide if life exists at all beyond on our own planet (mine's Earth, what's yours?). Second, one must determine how and where to look for it. You can apply statistical procedures to make both these decisions.

Estimating the Number of Smart Planets

In 1961, Frank Drake, an astronomer who was interested in looking at the universe from afar by reading radio waves (a bunch of which are bouncing off Earth all the time), decided to estimate how many other technologically advanced civilizations probably exist.

Being a little Milky Way–centric, he was most interested in determining the number of advanced worlds (planets willing and able to talk with us) that are nearby, in our own galaxy. Drake suggested this equation:

Number of Civilizations in the Milky Way $= (R)(N_h)(F_l)(F_i)(F_c)(L)$

Table 6-23 shows the meanings of the abbreviations in Drake's equation.

Table 6-23. Drake equation components

Term	Meaning
R	Rate at which new stars are produced in the galaxy (per year)
N_h	Average number of planets orbiting each star that can support life
F_l	Proportion of planets (from N_h) on which life does develop
F_i	Proportion of planets (from F_l) on which intelligent life develops
F_c	Proportion of planets (from F_i) on which civilizations develop
L	Average lifetime (in years) of civilizations (from F_c)

The formula is really nothing more than a chain of probabilities. The number of expected positive outcomes is determined by multiplying all the separate likelihoods together. Though a simpler equation without all the different permutations of *F* would work just as well, the specific different components were included to help scientists identify the important questions that needed to be answered to estimate the probability that we are not alone.

Applying Drake's Equation

To calculate a realistic number of planets in our galaxy that currently have intelligent life, you have to plug in some realistic numbers. Also, we know that the *correct* answer (the solution) must be at least 1, because there is intelligent life on Earth (insert your own joke here), and must be no more than 250,000,000,000 (the number of stars in the Milky Way) times the average number of planets around stars that could support life.

When the equation was first introduced, only one of the terms could be estimated with any consensus among astronomers. *R*, the number of new stars produced in our galaxy each year, is believed to be about 10.

If *R* were known to be 10 in the 1960s, I guess the correct number of stars in our galaxy would be closer to 250 billion + 40.

In 1980, Carl Sagan, popularizer of astronomy, discussed the Drake equation in his television series and book, *Cosmos*. Because we knew less about the planets in our own solar system then and, more importantly, knew nothing about planets in *other* solar systems (or even if there were such things),

Sagan's estimates for each value and his best-guess solution was somewhat speculative, but his answer was that about six million planets in the Milky Way at any given time have the technology to communicate with us.

Using what we know today, Table 6-24 provides one set of values that produces one possible answer. These values are taken from an essay in an October 2005 edition of *Astrobiology Magazine* (you probably have a copy on your coffee table) by Dr. Steven Soter of New York University. In some cases, I chose an exact value from Soter's discussion of a range of values.

Table 6-24. One application of Drake's equation

Term	Estimates	Calculations
R	10 per year	10
N_h	.01 (1 planet out of 100 stars)	$10 \times .01 = .10$
F_l	1 (assuming Earth is representative)	$.10 \times .10 = .10$
F_i	.001 (Soter suggests "small fraction")	$.10 \times .001 = .0001$
F_c	.20	$.0001 \times .20 = .00002$
L	100,000 years	$.00002 \times 100,000 = 2$

With these numbers, the equation estimates a total of two planets in the entire galaxy who could communicate with each other at any given time. Earth is one of those. What is the other?

As Sagan, Soter, and other authors point out, the number of planets in our galaxy that support advanced life at any given time depends on so many arbitrarily estimated factors that any little choice one makes when entering values dramatically changes the result. There is an important difference between six million possible friends and only two possible friends, but both estimates come from reasonable sets of assumptions.

Notice how the solution to the equation changes as you try different estimates for each component. If most groups of intelligent creatures—say, 80 percent, for example—eventually produce civilizations, the number of smart planets jumps to eight. If the average number of planets around a star that could support life is actually 2 (as Sagan suggests), our 8 would jump to 1,600 planets.

Soter advises that different reasonable estimates could produce an answer between a couple thousand and so few that our own planet's radio capabilities make it a statistical improbability, placing us as the only advanced civilization across many thousands of galaxies.

Finding our Space Chums

One possible outcome of the Drake equation is that there are only two planets in our galaxy with advanced intelligent civilizations capable of sending and receiving radio waves. If we really have only one other potential cosmic pen pal, it will be tough to find him or her or it in such a large haystack of planets. So, what to do?

The current strategy in seeking new life and new civilizations is to scan the skies with microwave receivers. Radio signals have a wide range of spectrums. Some occur naturally, and others are a particularly narrow range that are believed to only be created artificially—such as from the transmission of *Three's Company* TV episodes, or by radar, for example. By paying particular attention to those signals that are within this supposed artificial spectrum, those who search for alien life forms hope to discover and isolate either the random output of an advanced civilization or, perhaps, intentional signals broadcast for the benefit of any interested observer.

> If you own your own array of microwave listening stations, you'll want to tune them to the favored frequency for hunting life on other planets: 1.42 gigahertz. It is believed unlikely that any natural source would emit waves at that frequency.

The sky is big, though, and researchers use both targeted and convenience sampling techniques to decide where to look. The search strategy is to focus on a subpopulation of stars that meet two criteria:

- They are suns that share characteristics of our own.
- They are nearby (within a mere 100 light years of Earth).

Data Analysis

If the number of planets that could be emitting these key signals of life is very small (as some of the Drake equation permutations suggest), a search of this sample must be very thorough; otherwise, we might miss it. Statisticians would refer to this situation as a study that needs a great deal of power [Hack #8] because the *effect size* is so small.

There is so much data being collected as part of systematic efforts to scan the skies, no one person or even one computer can possibly analyze it all. You can help! SETI@home is a Berkeley University–based program that arranges for regular people with regular home or office computers to receive some of this data, so their computers can analyze it when they're not doing something else. *SETI* is the acronym for Search for Extraterrestrial Intelligence. The program works like a screensaver and can be downloaded for free at *http://setiathome.berkely.edu*.

The data won't make sense to you when you get it, but your computer will begin to use statistical analyses to sort through the signal information, looking for the telltale nonrandom narrow bandwidths that might mean another planet has reached the level of sophistication to produce something like *Gomer Pyle* or *Melrose Place*. You could be the first to discover life on other planets, so get to work!

Index

Symbol

! (exclamation point), 285

A

accuracy, concept of, 146
Aces
    counting, 174, 175
    Texas Hold 'Em and, 187
ACT (American College Test)
    standardizing scores, 111, 114
    test versions and, 134
    z scores and, 109
additive rule, 10, 12, 13
Adler, Joseph, 238–241
all-in, 162–164
analysis level (learning), 119, 120
analysis of answer options, 123–125
analysis of variance, 37
analytic view of probability, 14
answer options
    analysis of, 123–125
    multiple-choice questions, 117, 118
Apple iTunes, 224–229
application level (learning), 119–120
artificial intelligence, 216–220
averages, 88–91
axes, graphs and, 92–95

B

Bach, Johann Sebastian, 143
Bacon, Kevin, 308, 309, 313

bar bets
    designing, 194–197
    dice and, 184–186
    li'l flushes, 181–183
    matches in two card decks, 183
    sharing birthdays, 189
bar charts, 92, 93
base invariance, 273, 274
base rates, 147–149, 297
baseball games, 235–238
batting averages, 238–241
Bayer, Dave, 223, 224
Bayes, Thomas, 148, 149
Becker, T. J., 278
behavior, driving, 316–320
being-overtaken epochs, 317
Benford, Frank, 268, 278
Benford's law, 268–279
Bernoulli, Daniel, 204, 207
Bernoulli, Jakob, 18, 20
betting systems, 154–155
biased samples, 82, 142
big blind, 163
Big Slick, 188
binomial coefficient, 285, 287
binomial distribution, 201, 203
Binongo, J.N.G., 283
birthdays, sharing, 189–194
blackjack, 170–176
blinded out, 163
Bloom, Benjamin, 119, 121
Bloom's Taxonomy, 119–120
bottle-cap effect, 201, 203

We'd like to hear your suggestions for improving our indexes. Send email to *index@oreilly.com*.

breast cancer screening, 147–150
Browne, M., 278
Buffon's Needle Problem, 250–251
bust (blackjack), 170, 172
Butler, Bill, 215

C

Campbell, D.T., 35, 36
canonical variate analysis, 306
card games
    card-sharping, 186–189
    counting cards, 171, 174–176
    getting lucky, 181–184
    matches with two decks, 183
    probabilities and, 182
    rank ordering, 198, 199, 200
    shuffling cards for, 220–224
    wild cards, 197–200
    (see also Texas Hold 'Em)
card tricks, 220–224
card-sharping, 186–189
casinos
    card counting in, 171
    improving chances against, 174
    money and, 153, 154
    profit on roulette, 169
categorical measurement, 63–65
categorical variables, 66–70
cause-and-effect relationships
    correlation and, 46
    lottery numbers and, 181
    showing, 32–36
Central Limit Theorem
    beauty of, 97
    overview, 4–10
    t tests and, 72
central tendency, measures of, 88–91
chi-square test
    one-way, 60–66
    two-way, 65–70
ciphers, decoding, 300–304
classical test theory, 21–24, 136
cluster sampling, 83
coefficient alpha, 133, 135
coin toss
    heads or tails, 201–203
    Law of Total Probability, 190
    possible outcomes, 283–287
    probability of patterns, 264, 267
    St. Petersburg Paradox, 204–207

coincidences, interpreting, 259–263
collecting data, 15
Collins, Truman, 214
combinations, 265, 267
community cards
    flop as, 156
    improving hands and, 187, 188
    reading quickly, 189
comparison groups
    pretests and, 36
    t test, 70–74
comprehension level
    (learning), 119–120
CONCATENATE function, 241
concurrent validity, 139
conditional probabilities, 149
confidence intervals
    building, 23, 24
    Gott's Principle, 142, 144
    normal curve and, 105
    standard errors and, 77, 79
conjunction fallacy, 298
conjunction rule, 298
Conjunctionitus, 295–299
connections, 308–313
consequences-based arguments
    (validity), 138, 140, 141
constants, linear equations and, 49, 52
construct-based arguments
    (validity), 138, 139, 140
constructs, 281
content-based arguments
    (validity), 138, 139
contingency table analyses, 65–70
continuous values
    discrete values vs., 84–87
    graphs and, 92, 93
control groups, 34
convenience sampling, 83
Cook, T.D., 35, 36
Copernican Principle, 142
correct answers, 116, 123
correlation
    between variables, 236
    cause and effect and, 46
    defined, 38, 42
    direction of, 45
    effect size standards, 39
    factor analysis and, 280
    negative, 45, 47

Monty Hall problem, 209–211
multiple-choice questions
    analysis of answer options, 123
    writing good, 116, 117
multiple regression
    criterion variables and, 54
    defined, 230
    multiple predictor variables, 55–60
    predicting football games, 231–234
multiplicative rule, 10, 13
multivariate analysis of variance
    (MANOVA), 306
mutually exclusive outcomes, 190–191

N

negative correlation, 45, 47
negative numbers, 26
negative wording, 118
Newcomb, Simon, 268, 279
Nigrini, Mark, 268, 272, 275, 279
nominal level of measurement, 25, 27
non-experimental designs, 33
normal curve
    Central Limit Theorem and, 7
    overview, 97–100
    precision of, 23
    predicting with, 103–108
    z score and, 110, 111
normal distribution
    applying characteristics, 9
    iTunes shuffle and, 227
    overview, 97–100
    shape of, 9
    traffic patterns, 318
norm-referenced scoring
    defined, 101, 115
    percentile ranks, 107
    simplicity of, 111
null hypothesis
    defined, 16
    errors in testing, 31
    Law of Large Numbers and, 20
    possible outcomes, 16
    purpose, 17, 31
    research hypothesis and, 16
    statistical significance and, 31
nuts, 156, 161

O

observed score, 21, 22, 23
odds
    figuring out, 261, 262
    pot odds, 158–162, 179
    Powerball lottery, 177
    (see also gambling; probability)
one-way chi-square test, 60–66
ordering scores, 25
ordinal level of measurement, 25–27
O'Reilly Media, 145
outcomes
    blackjack, 172, 173
    coin toss, 283–287
    comparing number of possible, 260,
        261
    dice rolls, 184, 185
    gambler's fallacy about, 152
    identifying unexpected, 60–65
    likelihood of, 11–13, 157
    mutually exclusive, 190–191
    occurrence of specific, 194, 195
    predicting, 51–60, 264
    predicting baseball games, 235–238
    shuffled deck of cards, 223
    spotting random, 265–267
    trial-and-error learning, 216
    two-point conversion chart and, 242
outs, 158

P

pairs of cards, counting by, 176
parallel forms reliability, 132–135
partial correlations, 57
Party Shuffle (iTunes), 225, 227, 228
Pascal, Blaise, 285
Pascal's Triangle, 283–287
passing epochs, 317
payoffs
    expected, 205, 206
    magic number for lotteries, 179
    Powerball lottery, 177–179
Pearson correlation coefficient, 42–46,
    66
Pedrotti, J.T., 121
percentages
    ratio level of measurement, 26
    sample estimates, 75
    of scores, 99

Colophon

The tool appearing on the cover of *Statistics Hacks* is a Chinese abacus, or *suanpan*. Centuries before the emergence of the written Hindu-Arabic numeral system, the abacus, often constructed of a wooden frame with beads sliding on wires, was used as a calculation tool. Historians place its invention between 2,400 and 300 BC. At that time, when most people could not read or write, it might have seemed ridiculous to scribble symbols on expensive papyrus when such an excellent calculating device was available. The *suanpan* differs from the European abacus in that its board is split into two parts. The lower part holds five counters on each wire; the upper section holds two. Complex *suanpan* techniques accomplish not only simple addition, but also multiplication, division, subtraction, and square and cube root operations efficiently.

The cover image is a stock photograph from CMCD Everyday Objects. The cover font is Adobe ITC Garamond. The text font is Linotype Birka; the heading font is Adobe Helvetica Neue Condensed; and the code font is LucasFont's TheSans Mono Condensed.

Better than e-books

Buy *Statistics Hacks* and access the
digital edition FREE on Safari for 45 days.

Go to www.oreilly.com/go/safarienabled
and type in coupon code 5NG4-7III-9MQW-N2IB-GXCL

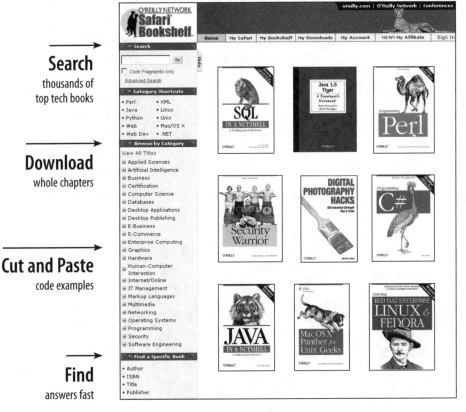

<table>
<tr><td>→ **Search**
thousands of
top tech books</td></tr>
<tr><td>→ **Download**
whole chapters</td></tr>
<tr><td>→ **Cut and Paste**
code examples</td></tr>
<tr><td>→ **Find**
answers fast</td></tr>
</table>

Search Safari! The premier electronic reference
library for programmers and IT professionals.

Related Titles from O'Reilly

Hacks

Access Hacks

Ajax Hacks

Amazon Hacks

Astronomy Hacks

Baseball Hacks

Blackberry Hacks

BSD Hacks

Car PC Hacks

Digital Photography Hacks

Digital Video Hacks

eBay Hacks, *2nd Edition*

Excel Hacks

Flash Hacks

Flickr Hacks

Firefox Hacks

Gaming Hacks

Google Hacks, *2nd Edition*

Google Map Hacks

Greasemonkey Hacks

Halo 2 Hacks

Hardware Hacking Projects for Geeks

Home Theater Hacks

iPod & iTunes Hacks

IRC Hacks

Knoppix Hacks

Linux Desktop Hacks

Linux Multimedia Hacks

Linux Server Hacks

Linux Server Hacks, Volume 2

Mac OS X Panther Hacks

Mapping Hacks

Mind Hacks

Mind Performance Hacks

Network Security Hacks

Nokia Smartphone Hacks

Online Investing Hacks

Palm & Treo Hacks

PayPal Hacks

PDF Hacks

PC Hacks

PHP Hacks

Podcasting Hacks

PSP Hacks

Retro Gaming Hacks

Skype Hacks

Smart Home Hacks

Spidering Hacks

Swing Hacks

TiVo Hacks

Visual Studio Hacks

VoIP Hacks

Web Site Measurement Hacks

Windows Server Hacks

Windows XP Hacks, *2nd Edition*

Wireless Hacks, *2nd Edition*

Word Hacks

XML Hacks

Yahoo! Hacks

O'REILLY®

Our books are available at most retail and online bookstores.

To order direct: 1-800-998-9938 • *order@oreilly.com* • *www.oreilly.com*

Online editions of most O'Reilly titles are available by subscription at *safari.oreilly.com*